THE STORY OF LANKA ELECTRICITY COMPANY

JULY 2022

ADB

ASIAN DEVELOPMENT BANK

About the Serendipity Knowledge Program

The Serendipity Knowledge Program is an ADB platform dedicated to identifying knowledge solutions for
Sri Lanka's development challenges. Serendib is one of Sri Lanka's ancient names and serendipity refers to
a fortunate finding, which is a common occurrence throughout the history of product invention and scientific discovery.
ADB established this new knowledge program in 2021 in line with its vision as a knowledge solutions bank.

Contents

Figures and Table   v

Foreword by Secretary of Ministry of Power, Government of Sri Lanka   vi

Foreword by Director General, South Asia Department, ADB   vii

The Story of Lanka Electricity Company (Pvt) Limited   viii

Abbreviations   ix

Introduction   1
    ADB and LECO   1
    World at the Time   2
    Utility Reforms in South Asia   2
    ADB Energy Sector Operations   3

History of the Electricity Supply Industry   5
    Council-Managed Distribution Networks   5
    Role of Ceylon Electricity Board: The National Utility   7

The Birth of Lanka Electricity Company   11
    Political Leadership   11
    The Objective   12
    Birth of Lanka Electricity Company   13
    Ownership of the New Company   14
    A New Company and New Thinking   14

Teething Problems of the New Distribution Utility   21
    Taking Over Assets and Service Areas   21
    Mapping before Taking Over   23
    Asset Valuation   24

A New Approach to Engineering and Customer Care   **33**
Administrative Independence   33
Distribution Planning: Key to a Successful Utility   35
Treasures from the Archives   38

The Success Story in Electric Utility Management   **46**
How Was It Done?   46
From a Loss-Making Start-Up to a Profit-Making Utility   47
LECO Operations 2020   52

Gearing up for the Future   **54**
The Company Ventures into Other Businesses   54
Support to Accelerate Deployment of Rooftop Solar Power   55
Solving Emerging Problems of Distributed Generation   56
Upgrades for Improved Reliability   57
The Next Move: Distributed Generation in Smart Grids   59
Financial Independence   60
Past and Emerging Challenges   61

References   **65**

Figures and Table

Figures

1   Evolution of Electricity Supply Industry Worldwide   2
2   Energy Losses in the LECO Network   46
3   Revenue Flow in Sri Lanka Electricity Industry   49
4   Allowed Profits and Actual Profits of LECO's Distribution and Supply Activities   51
5   Planned Upgrades to the LECO Network in Kelaniya   58

Table

Allowed Revenue for LECO's Distribution and Retail Supply   50

Foreword by Secretary of Ministry of Power, Government of Sri Lanka

Lanka Electricity Company (Pvt) Limited (LECO) is considered one of the most important achievements in the Sri Lanka electricity supply industry. The visionary leadership and the efforts of those who believed in much-needed operational efficiency improvements and customer care in electricity distribution, created and developed LECO to become the leader in innovation since its inception in 1983. The pioneering efforts of LECO led to many technical and other interventions as far back as the 1980s, which drastically reduced the technical and commercial losses, increased reliability of supply, and improved utility management and customer services.

This publication on LECO clearly demonstrates different aspects of utility development, challenges, and how LECO has succeeded in overcoming those challenges over the years. These lessons from LECO will surely be excellent examples to be followed in Sri Lanka and outside.

I am extremely happy to see the Asian Development Bank's close and continuous involvement and support in the development of LECO and the electricity industry in general, as the lead multilateral development partner in the Sri Lanka power sector.

Wasantha Perera
Secretary
Ministry of Power
Government of Sri Lanka
April 2022

Foreword by Director General, South Asia Department, ADB

Lanka Electricity Company (Pvt) Limited (LECO) is one of the earliest examples of the recent phase of electricity industry unbundling and reforms. The Government of Sri Lanka faced the problem of underperforming electricity distribution by municipal, urban, and town councils. Instead of burdening the national utility Ceylon Electricity Board, which was the natural choice to absorb the poorly run council networks, the government decided to set up LECO.

The Asian Development Bank (ADB) has been closely involved with the development of LECO over its formative years. Technical assistance provided by ADB, blended with the talents of LECO's well-qualified and motivated staff, enabled three projects to be implemented during the company's first decade of operations. Developing an organizational culture that prioritizes planning, distribution engineering, customer care, and financial viability, LECO weathered many a storm in the 1980s and 1990s.

LECO truly is an example to the utility industry in South Asia and beyond.

More recently, ADB has worked with LECO to pilot a smart metering program and a renewable energy microgrid, contributing to development of analytical skills and research potential in LECO and Sri Lanka's universities.

This is the "Story of LECO," dedicated to all who worked toward its goals from within and outside LECO.

Kenichi Yokoyama
Director General
South Asia Department
Asian Development Bank

The Story of Lanka Electricity Company (Pvt) Limited

Lead Authors:
- Tilak Siyambalapitiya
- Priyantha Wijayatunga
- Ranishka Wimalasena

Contributing Authors:
- W A L W Amarasiri Perera
- Ramani Nissanka
- Swetha Perera
- Muditha Karunathilake
- Nadeesha Chandramali Perera
- Tharindu de Silva
- Dasun Ranasinghe
- Narendra de Silva

Acknowledgments:
- Jayasiri Karunanayake, Shanthi Amaratunga, G.A.D. Sirimal, Sanath Kumara, *for reviewing the manuscript*
- Sumadhurie Hansika *for technical review and editorial assistance*
- Madhura Gamage of LECO *for insights into regulatory compliance*
- W.A.L.W. Amarasiri Perera and G.A.D. Sirimal *for providing archives related to LECO's formative years*
- Tharindu De Silva, Thisara Karunathilake, M. A. Pushpa Kumara, Duranka Menerigama, Bhasara Sirisinghe, and Harsha Wickramasinghe *for photographs*
- All employees of LECO for providing information and archives
- Staff of Sri Lanka Resident Mission and Energy Division of South Asia Department, Asian Development Bank for overall support

Reviewed by:
- Shavindranath Fernando, general manager, Ceylon Electricity Board and Board Member, LECO 2013–2015

Abbreviations

ADB	Asian Development Bank
CEB	Ceylon Electricity Board
GIS	geographic information system
GWh	gigawatt-hour
km	kilometer
kV	kilovolt
kVA	kilovolt-ampere
kWh	kilowatt-hour
LECO	Lanka Electricity Company (Pvt) Limited
MW	megawatt
PV	photovoltaic
SLRs	Sri Lanka rupees
UDA	Urban Development Authority
V	volt

Introduction

ADB and LECO

Improving governance has always been key in the development agenda of the Asian Development Bank (ADB) for energy sector operations. These intentions were evident even before the worldwide wave of power sector reforms were implemented in the 1980s, and in South Asia in the mid-1990s.

ADB in Sri Lanka supported power sector reforms even in the early 1980s. Sri Lanka's reforms in power distribution commenced in 1983 with these initiatives.

The initial support ADB provided in the creation of Lanka Electricity Company (Pvt) Limited (LECO) and its sustenance over the years are among the efforts to improve governance in Sri Lanka's electricity distribution sector, which eventually led to significant improvements in performance. Looking at LECO's continuous improvement and ability to set new performance benchmarks in the country's distribution sector, ADB is proud of this creation and its association in the process.

This publication, *The Story of Lanka Electricity Company,* presents the landmark achievements of LECO, then a new distribution utility established in Sri Lanka in 1983.

LECO file photo

World at the Time

Power sector structures and institutions across the world have gone through many phases of reforms, from private sector-driven systems to completely nationalized state-owned systems, and since 1980s (see figure 1), back again to systems with significant levels of private participation alongside state-run systems.

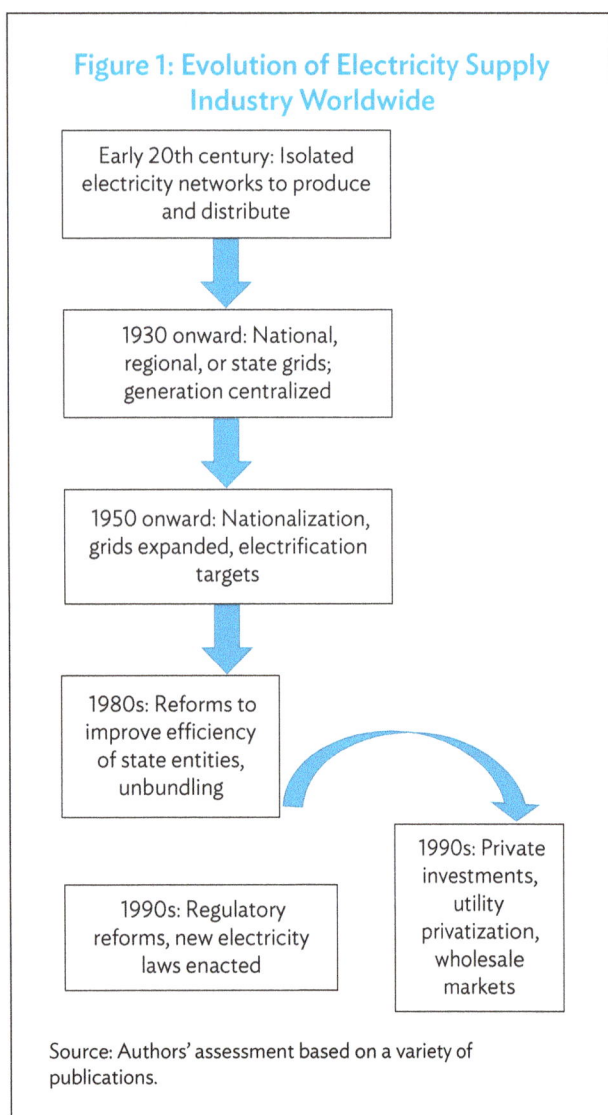

Figure 1: Evolution of Electricity Supply Industry Worldwide

Early 20th century: Isolated electricity networks to produce and distribute

↓

1930 onward: National, regional, or state grids; generation centralized

↓

1950 onward: Nationalization, grids expanded, electrification targets

↓

1980s: Reforms to improve efficiency of state entities, unbundling

1990s: Regulatory reforms, new electricity laws enacted

1990s: Private investments, utility privatization, wholesale markets

Source: Authors' assessment based on a variety of publications.

The wave of reforms that commenced in the 1980s initially caused countries such as Chile to introduce independent regulation. Vertical and horizontal unbundling of utilities, establishing wholesale and retail markets, and pricing reforms followed.

Most countries in the developing world implemented these reforms mainly at the insistence of international financing institutions, backed by national governments with neoliberal economic policies.[1]

Utility Reforms in South Asia

The wave of reforms sweeping South America and parts of Europe in the 1980s reached South Asia only about a decade later in the mid-1990s, commencing with reforms in the Indian power sector initiated by the World Bank.

The creation of LECO in Sri Lanka, however, was in 1983, a decade before widespread reforms reached South Asia.

At the time of establishing LECO, the power sector in South Asia was managed largely by state-owned utilities. Except in a few urban areas of India and some community-owned smaller distribution systems in rural areas, all distribution operations served by the national grid were state-owned. Utilities were government departments, authorities, or boards, with very little financial autonomy or administrative independence. In many cases, these distribution systems were owned and operated by poorly equipped local institutions or authorities, such as city councils.

Access to the national electricity grid was extremely low, in the range of 10%–30% of the population in South Asia. The situation in rural areas was even worse.

---

[1] A. Lee and Z. Usman. 2018. Taking Stock of the Political Economy of Power Sector Reforms in Developing Countries. *World Bank Group Policy Research Working Paper* No. 8518. Washington, DC: World Bank.

Load shedding, blackouts, and brownouts were frequent throughout South Asia.

Overall, lack of investments to increase access; absence of the mindset for supply reliability; noncompliance with internationally accepted technical standards in planning, construction, and operation; and poor customer service were the order of the day in electricity distribution in South Asia.

ADB Energy Sector Operations

The energy sector operations in the late 1980s and 1990s were mainly framed around strategic objectives of economic growth, poverty reduction, human resource development, and environmental protection. Along with these objectives and political upheavals in the Gulf in the 1990s led ADB and its developing member countries to focus on energy source diversification to strengthen energy security. Rising energy prices required emphasis on energy efficiency (on both the supply side and the demand side), energy conservation, and more effective energy planning.

Furthermore, in the 1990s, ADB commenced focusing on integrated packages of development,

consisting not only of project financing, but of softer interventions as well. These integrated packages included assistance with policy reform; capacity development; and regional cooperation based on strategic objectives of economic growth, poverty reduction, human development, improvement of status of women, and environmental protection.

Against this backdrop, LECO was established with the intention of improving operational efficiency in power distribution in Sri Lanka, mainly focusing on the geographic areas supplied by local authorities such as municipal, urban, town, and village councils.

This exercise entailed complete rehabilitation of such distribution systems with the introduction of new technology and system designs, visible improvement of service standards including customer care, and most importantly, a management that was independent of direct political interference. To date, LECO continues to progress in all these areas since its inception.

The story of LECO is a presentation of the problems LECO set out to resolve, its achievements, constraints, and an insight into emerging challenges and the company's future directions.

History of the Electricity Supply Industry

Sri Lanka's population in 1970 was 12.5 million, but only a dismal 8% of households had electricity. Benefits of hydropower development since 1950 and the growing investments in the transmission network were not reaching the public at large.

In 1970, over 90% of the population used kerosene for lighting at home.

Large hydroelectric power plants were developed at a brisk speed and by 1980, 330 megawatts (MW) of hydropower were in operation. Another 660 MW of hydropower plants were under construction. In certain years, the country reported a surplus in

Wimalasurendra hydropower plant. One of the 15 hydropower plants built over 1950–1990 that enabled Sri Lanka to be a 95% renewable energy power system by 1995 (Photo by Uthpala Sumithraarachchi, CEB).

generation, and the government was encouraging industrial customers to use more electricity from the grid. Despite the increased generation, even by 1980, only 12% of households had electricity.

The transmission network development commenced in 1950, initially at 66 kilovolt (kV), and later at 132 kV and 220 kV.

The country's electricity networks needed new investments on two fronts—one to expand the transmission and distribution network to reach the widely dispersed households and small businesses, and the other to upgrade the aging distribution networks dating from the 1930s that were historically managed by municipal, urban, town, and village councils.

Council-Managed Distribution Networks

Even before the national grid was established in 1950, certain councils (mostly municipalities) provided electricity supply to city dwellers using the councils' own diesel generators. In some towns, electricity was provided for only a few hours in the evening. When the national grid arrived in the 1950s, diesel generators were disconnected, and grid electricity was purchased by the councils to serve their customers throughout the day. However, most municipalities maintained the diesel generators for some time to be used for peak lopping, in emergencies, and in times of power curtailments in the grid.

Despite cheaper hydroelectricity arriving at their doorsteps since 1950, not everything was smooth nor proper in these council-managed distribution networks. The growing management crisis of distribution networks was apparent in the distribution losses (estimated) that stood at 25% of energy in 1970, which then grew to 28% by 1980.

> Over one-quarter of electrical energy delivered to a council network was lost because distribution lines and transformers were weak, metering was bad, and pilferage was rampant.

In 1980, Sri Lanka reported the national transmission and distribution losses in council-served areas and elsewhere, to be 20% of net electricity generation.

Major cities such as Kandy, Galle, Jaffna, Matara, Negombo, Moratuwa, and Kotte each had more than 5,000 customers connected in their respective council-managed distribution networks.

Old electricity meters. The metering system needed revamping (Photo by Nilush Jayawardana, CEB).

The capital city of Colombo was served first by a company, Bousted Brothers, thereafter by the Department of Government Electrical Undertakings, and since 1969, by its successor, Ceylon Electricity Board (CEB), the national utility.

Transformers in 1970. Old transformers were badly in need of replacement and technology upgrades (Photo by Nilush Jayawardana, CEB).

Role of Ceylon Electricity Board: The National Utility

By 1980, about 25% of the country's electricity distribution network was managed by councils and the rest was the responsibility of CEB, the national utility. Specific large customers within council areas were also served by CEB.

The sales composition of CEB was 50% to industries, 25% to councils, and the balance to CEB's own retail customers. CEB's transmission and distribution losses were about 11%, much lower than that in council networks.

Customer tariffs of each council network were determined by the chief electrical inspector, a position established under the Electricity Act. As a result, customer tariffs across the councils were different and such determinations rarely followed a

Galle ~ 1980

A town in Sri Lanka in 1980. Facilities in towns were limited in the 1980s, aggravated by limited availability and poor quality of electricity supply (File photo).

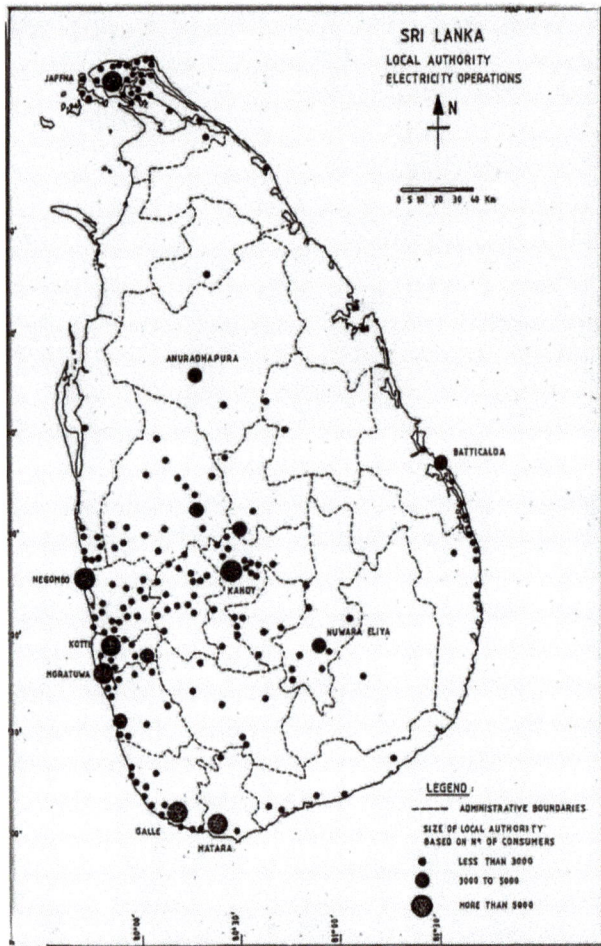

SRI LANKA
LOCAL AUTHORITY
ELECTRICITY OPERATIONS

Council-owned distribution networks. By 1982, there were 219 council networks in different geographic locations, and almost all were facing technical and financial difficulties. Source: TA 578-SRI, Final Report October 1984, ADB.

Dilapidated networks prevented councils from providing electricity connections to large commercial and industrial customers, and CEB was compelled to make inroads into council-managed areas to provide electricity to them.

waiting list to obtain electricity connections to new dwellings. The distribution networks within the council required upgrading to provide additional connections. For most would-be customers, the network upgrade either materialized very late or never occurred.

People had to wait for years to receive a connection and then suffer from the poor quality of supply. Night-time brownouts and day-long blackouts were the order of the day in council networks in the 1970s.

"Living in the city of Negombo, I waited for 1 year to receive an electricity connection to my new house built in the 1980s. The service quality was poor, unreliable," recalled A. C. B. Fonseka, an executive working in the nearby export processing zone.

systematic approach to provide a reasonable revenue to councils to manage and develop their networks.

Poor technical and financial management, inadequate tariffs, and confining their electricity business mainly to low-yielding retail customers caused councils to become cash-strapped. The required investments to expand and upgrade their distribution networks never materialized.

The quality of service was declining, and customers had to register themselves on a

It is argued that distribution networks managed by councils or cooperatives were the way to go in the 1970s and the 1980s, which met with success in Europe, Asia, and the United States. Some countries still operate distribution networks owned and operated by councils or cooperatives.

However, constrained by the absence and inability to retain quality technical and managerial staff, poor financial management, and incorrect electricity pricing, the councils in Sri Lanka could not manage their networks. Billing for electricity use was erratic and collection rates were poor in many councils.

The first reaction to the cash crisis in any council was not to pay CEB, the bulk supplier, for the energy purchased. With nearly 25% of its sales not being paid for, along with its own financial constraints, CEB was feeling the impact of the poorly managed council networks.

ලංකා විදුලිබල මණ්ඩලය
இலங்கை மின்சார சபை
CEYLON ELECTRICITY BOARD

Many households without grid supply had stand-alone solar units since 1970s (LECO file photo).

The Birth of Lanka Electricity Company

Political Leadership

The electricity supply was rapidly deteriorating. The supply voltage in some council-owned networks was as low as 110-volt during the evening peak period, against the standard distribution voltage of 230 volts. Some households and shops were believed to keep and use several 110-volt lamps during such voltage drops. These overloaded distribution lines wasted about 50% of its input energy.

The President of the country, also the Minister of Power and Energy by 1982, suggested that CEB, the national generation, transmission, and distribution utility, should take over the distribution networks managed by the councils.

However, CEB was reluctant to accept the President's suggestion because, among other reasons, inefficiency was widespread among officials and technicians in the councils. CEB viewed that such staff joining from councils would spoil the discipline of CEB's own staff. CEB was alarmed with the poor financial status of the council networks, poor revenue collection, and network losses of almost 50%, compared to CEB's own transmission and distribution losses of only about 17%.

Public complaints about the poor quality of electricity supply in council areas were too many for the President to ignore. Besides, much of the country's social and economic activities occurred within these council-served areas.

Finally, the President appointed a committee in 1982 to recommend a suitable action plan to solve many problems faced by customers. The committee was chaired by K. K. Y. W. Perera, who was the secretary of Ministry of Power and Energy, and the committee consisted of four other members including H. S. Subasinghe, the general manager of CEB.

The committee, with the available data of 219 council-owned networks serving 230,000 customers, analyzed a number of options, ranging from handing over the services to CEB to complete privatization. The committee submitted its recommendation, in November 1982, to form a company with shareholding from CEB, Urban Development Authority (UDA), and councils.

However, there was significant resistance from the councils to the move. With resistance mounting, the government invited all the council chairperson/s to a seminar in the capital Colombo. Officials explained the need for reforms and that a company could serve the purpose. The councils would be shareholders by virtue of their electrical assets handed over to the company. The President attended the lunch with the chairperson/s of councils and expressed his determination to proceed. By the end of the seminar, all chairperson/s, in principle, agreed with the proposal.

The Objective

The committee, appointed in 1983, analyzed the situation and identified main objectives of the reorganization.

Objectives

1. Improvement of quality of electricity supply.
2. Reduction of technical and commercial losses.
3. Improvement of billing, revenue collection, and payment to CEB for bulk purchases.
4. Minimization of disparities in retail prices of electricity in different areas.
5. Improvement in courtesy to public and attention to complaints.
6. Improvement of the quality of management and staff.[2]

To achieve the objectives, the committee in 1982, analyzed six options for future ownership and management.

1. **Councils continue to handle their own electricity distribution systems (a "do nothing" scenario).**

Only a few councils distributed electricity with a satisfactory quality. On the national level, the quality of service was not acceptable. The potential for the service to be improved within the existing arrangements was minimum, due to the weak management structure of councils. Therefore, this was not a suitable option to achieve the defined objectives.

2. **A "one-day" takeover by CEB; or**

3. **A "staggered" takeover by CEB.**

CEB, the bulk supplier to councils, was itself facing a crisis owing to the large-scale exodus of its qualified engineers and other technical staff. CEB had already undertaken a large (SLRs5 billion) capital development program on its assets. Therefore, CEB taking over dilapidated council networks and thereby doubling its customer base, was not an option to achieve the defined objectives.

4. **Formation of an Electricity Distribution Board.**

The formation of a separate Electricity Distribution Board offered several advantages. However, due to possible funding problems, difficulties in recruitment and retention of quality staff, and the possible delays associated with required new legislations were considered against this option.

5. **Formation of an Electricity Distribution Company (private sector).**

A private company would have the potential to improve activities of electricity distribution, collection of revenue, improvement of electricity supply quality, prompt attention and courteous response to public, recruitment of qualified staff, against any other option. However, a mindset of a private company to focus on higher profits and to lose the broader considerations of the government, and weak (or nonexistent) regulatory framework to effectively facilitate a private utility weighed against this option.

6. **Formation of an Electricity Distribution Company (with CEB, UDA, and council participation).**

A distribution-oriented company could focus its attention solely on problems associated with electricity distribution and customer care. This company would be able to attract quality staff and manage them efficiently, owing to the flexibility in a company environment. Improvements could be implemented without external interference or financial constraints.

Minimizing or eliminating disparities in retail prices across council networks would also be possible.

---

[2] Government of Sri Lanka, Ministry of Power and Energy. 1982. Report of the Committee appointed by His Excellency the President to recommend a suitable scheme for the improvement of electricity distribution in Local Authority areas, November.

A company with the majority holdings of CEB, with the participation of UDA and councils, could propagate broader policies in electricity tariffs and through its shareholdings, would reflect the objectives of the government.

Birth of Lanka Electricity Company

Considering all the facts, the committee recommended to formulate a limited liability company, floated jointly by CEB, UDA, and the council's participation to take over electricity distribution in council areas. The assessment of options was qualitative, as no information or precedents were available in similar environments even in other countries. The government swiftly approved the committee's recommendations. The Cabinet of Ministers approved the formation of an electricity distribution company. Consequently, LECO was formally incorporated on 19 September 1983 under the provisions of the Companies Act No. 17 of 1982.

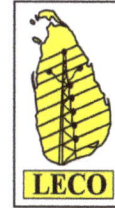

LECO

සීමාසහිත ලංකා විදුලි (පුද්ගලික) සමාගම
வரையறுக்கப்பட்ட இலங்கை மின்சார (தனியார்) நிறுவனம்
LANKA ELECTRICITY COMPANY (PVT) LTD

Hope for a new company and better services (Photo by M. A. Pushpa Kumara).

With the initial incorporation and establishment of LECO, the government had to handpick the key persons to lead this unique effort to restructure and reform electricity distribution.

Ownership of the New Company

LECO's articles of association allows a minimum of five and a maximum of seven directors. CEB was entitled to appoint five directors, of whom two would be ex officio: the chairman/men and the general manager of CEB. UDA was entitled to appoint two directors, one being the commissioner of local government. The first board of directors of the company reflected the importance the government placed on the company. All members were highly experienced in their professions; four out of six directors had decades of electric utility experience.

The authorized capital of the company was Sri Lanka rupee (SLRs) 1,200 million divided into 120 million shares of SLRs10 each. There were two types of shares: 75% of the total shares were type A shares with voting rights, and the remaining 25% shares were type B shares without voting rights. As a private company, the number of members was limited to 50 and shares were not issued to the public. CEB was to hold a minimum of 51% of type A shares. The shares issued to the councils that opted to receive shares, equivalent to the value of assets transferred, were limited to type B shares, without voting rights.

In accordance with the provisions of the then Electricity Act No. 19 of 1950, the license to supply electricity was issued to LECO on 25 May 1984. The license was in force for a period of 5 years and provided the authority to supply electricity anywhere in the country, without restricting to a specific geographic area, reflecting the planned expansion of the LECO service area. The voltage and the system of supply as authorized in the license were 33 kV, 11 kV, 400/230 V, 50 hertz, alternating current of 2, 3, or 4 wire system.

A New Company and New Thinking

LECO swiftly commenced operations in 1983 itself, initially by setting up an office in a rented house in Colombo, with a limited staff. LECO handpicked and hired engineers with local and international experience in power distribution to lead planning, projects, and operations.

There was a transparent procedure to hire staff. Although LECO was owned indirectly by the government, most importantly, it was not required to abide by the practices followed by other mainstream government agencies in recruitment, human resource management, and development.

LECO believed in policies and principles of distribution planning and meticulously practiced them. LECO knew exactly to which medium voltage feeder, distribution transformer, and low-voltage line each customer was connected to. Load flow studies were conducted for the entire distribution network at a time when the belief was such analyses were required only in transmission planning.

Any successful company must efficiently collect their revenue to remain in business and curb losses. An innovation brought in by LECO from its inception was instant billing. Immediately after reading a meter, the bill was produced and delivered to the customer's doorstep. By then, although payment at banks was allowed, customers had to fill forms or slips. LECO's bill included a section to be used as a paying-in slip which did not require the customer to fill up additional information. Additionally, LECO provided the option for customers to pay their bills at local shops.

LECO was instrumental in minimizing commercial losses. A meter was fixed at the low-voltage end of each distribution transformer. Every transformer was metered, and the readings were checked against aggregate customer bills, on a monthly basis. Any discrepancy indicative of theft or metering problems received immediate attention.

The formation of LECO broke away the traditions in utility engineering and management and showed that even a government-owned entity can be efficient when the right leadership and freedom to make decisions are granted. The people—from chairman/men down to the line crew—were LECO's assets, who facilitated this transformation.

The Memorandum of Association states LECO's primary objective is "To carry out in Sri Lanka the business of maintenance, improvement, supply, development, expansion, generation, distribution, and sale of electrical energy."

Reading the meter. LECO staff reading the meter on a roadside distribution transformer, a practice since 1980s that helped monitor losses (Photo by Tharindu de Silva, LECO).

Synergy. Overcoming the need and the limitations in congested urban areas (LECO file photo).

Long-Term Vision for the Power Sector Starting with Distribution Reforms

K. K. Y. W. Perera recalls his stint as the Chairman/men of the Committee on Electricity Distribution Reforms

In 1981, K. K. Y. W. Perera, an electrical engineering graduate of the then University of Ceylon, an academic of University of Moratuwa, Sri Lanka, was seconded to lead the energy sector as secretary, Ministry of Power and Energy. For over a decade, among his other responsibilities, he led the formation of LECO, first as chairman/men of the government committee that recommended the creation of LECO and then as a member of its first board of directors. He went on to oversee LECO's formation and subsequent technical and financial performance. He was also a practicing engineer in the Department of Government Electrical Undertakings for over 10 years. Perera is currently the chancellor, University of Moratuwa, Sri Lanka.

Perera went through years of his stint with LECO. He explained the status of electricity supply at the time of LECO's establishment. The supply voltage in some council-owned networks was as low as 110 V during the evening peak period, against the nominal distribution voltage of 230 V. Apparently, even some households and shops kept several 110 V lamps and used them during such voltage drops. These overloaded distribution lines wasted about 50% of its input energy. Also, it was difficult to get new electricity supply connections because the councils refused, using the excuse that their transformers were already overloaded.

Sri Lanka's President at the time, also the Minister of Power and Energy, suggested that CEB, the national generation, transmission, and distribution utility, should take over the distribution networks managed by the councils. However, CEB was reluctant to accept the President's suggestion because, among others, corruption was widespread among some officials and technicians managing such council-owned networks. CEB viewed that such staff joining from councils would spoil the discipline of CEB's own staff. CEB was alarmed with the poor financial status of the council networks, poor revenue collection, and network loss of almost 50%, at a time when its own transmission and distribution losses were only about 17%.

The President requested that Perera examines this issue as a national problem and to come up with a solution. For this purpose, a committee chaired by Perera as secretary to the Ministry was appointed with R. Abeyratne (additional secretary, Local Government), H. S. Subasinghe (general manager, CEB), M. Somasunderam (director, Public Enterprise, Treasury), and P. B. N. Fernando (deputy general manager, CEB) as members. G. A. D. Sirimal (assistant secretary, Power and Energy) assisted the committee. Armed with the data for 219 council-owned networks serving about 230,000 customers at the time, the committee analyzed a number of options, ranging from handing over to CEB up to complete privatization. The recommendation of the committee was to form a company with shareholding from CEB and UDA, and councils. The committee submitted their recommendations in November 1982, which the government approved.

However, there was significant resistance by the councils to the move. With resistance mounting, the government invited all the council chairmen to a seminar in Colombo. Officials explained the need for reforms and that LECO could do this. The councils would be shareholders by virtue of their electrical assets handed over to LECO. The President attended the lunch with the council chairmen and expressed his determination to proceed. By the end of the seminar, all chairmen agreed in principle.

LECO commenced operations in late-1983, initially at a rented house in Colombo, with a limited staff. LECO handpicked and hired engineers with local and international experience in power distribution to lead planning, projects, and operations. There was a transparent procedure to hire staff. Most importantly, although government-owned, LECO was established under the Companies Act and was not required to abide by the practices followed by other mainstream government agencies, including in human resource management and development.

LECO believed in and meticulously practiced the policies and principles of distribution planning. As far back as 1990, LECO knew exactly to which medium voltage feeder, distribution transformer, and low-voltage line each customer was connected to. Load flow studies were conducted for the entire distribution network at a time when the belief was such analyses were required only in transmission planning. LECO followed the policy of minimum inconvenience and harm to people living beside the lines and trees growing under the lines. LECO locally designed high-technology poles which do not need stay wires or struts to support lines at curves. A distribution transformer could be mounted on a standard pole, saving ground space. An eminent local engineer, M. Chandrasena, designed these poles and a local manufacturer built them to meet stringent standards. These poles were taller than the usual poles to allow more ground clearance and to minimize damage to taller trees. LECO decided that only aerial bundled conductors would be used for low-voltage lines, to ensure safety and minimum damage to flora and fauna. That was in 1984.

Another innovation brought in by LECO, right from its inception, was instant billing. Immediately after reading a meter, the bill was issued and delivered to the customer's doorstep. LECO's bill included a section to be used as a paying-in slip which the customer did not have to fill up additionally and also gave the option to customers to pay the bills at selected shops. These measures resulted in shorter revenue collection periods and better collection rates. Furthermore, LECO metered the output of each distribution transformer to cross-check against aggregated customer bills, on a monthly basis. Any discrepancy owing to theft or metering problems received immediate attention.

The other interesting intervention at the time was that a maintenance crew did not include anyone who was not directly involved in the maintenance work. For instance, there were no separate drivers for the vehicles.

Looking back, Perera was proud and content to have provided leadership to establish a completely new culture in power distribution in Sri Lanka, breaking away from traditional utility engineering and management, even under state ownership. LECO has now grown into an institution full of talent and continued innovation, with remarkably low levels of distribution losses, currently standing at less than 4%.

LECO needed to be led by mature, talented leaders with innovation

Mohan Munasinghe, Senior Energy Advisor to the President 1982–1986, recalls how distribution reforms commenced in Sri Lanka

Mohan Munasinghe is Chairman/men, Munasinghe Institute for Development (MIND) and MIND Group, and distinguished guest professor, Peking University, People's Republic of China. He holds postgraduate degrees in electrical engineering, physics, and economics from Cambridge University, United Kingdom (UK); Massachusetts Institute of Technology, United States (US); and McGill and Concordia Universities, Canada. He has been a senior advisor to the presidents of three countries, including serving as senior energy advisor to the President of Sri Lanka, 1982–1986. He was division chief for Energy, Infrastructure, and Environmental Policy (1987–1995), and senior advisor for Sustainable Development (1996–2002) at the World Bank in Washington, DC.

His research awards include sharing the 2007 Nobel Peace Prize (as vice chair, Intergovernmental Panel on Climate Change) and being the 2021 Blue Planet Prize Laureate. He has written over 120 books and 350 technical papers.

"In the early 1980s, when I commenced my job as senior energy advisor to President J. R. Jayewardene, increasing the efficiency of state institutions and reducing losses was a high government priority," said Mohan Munasinghe. This was consonant with the World Bank's ongoing emphasis on privatization to increase efficiency. Among the energy sector institutions targeted for early reform were local authorities such as municipal councils and town councils that distributed electricity in many urban areas, because of their high technical and financial losses.

Munasinghe worked with K. K. Y. W. Perera, secretary, Power and Energy, to set up LECO as an umbrella institution to take over council-run networks. Sri Lanka was among the first developing countries to implement such reforms. The vision was not to privatize state bodies, but to create an institution that was government-owned, while functioning as efficiently as a private entity. The unique features of LECO were its public ownership, combined with a high degree of autonomy and professional leadership. This prevented government interference that was often the root cause of inefficiency and corruption.

The success story of LECO begins with the appointment of H. S. Subasinghe as the first chairman/men to lead this novel effort. As highlighted in Munasinghe's memo to President Jayewardene in March 1983, General Manager Subasinghe of CEB the national utility, was about to retire, and the country could ill-afford to lose his considerable skills, integrity, and honesty.

The President agreed with Munasinghe's assessment that Subasinghe should be earmarked as chairman/men of LECO, and be given an early mandate to help design the legislation, functions, organizational framework, staffing, and other matters. Subasinghe, Perera, and Munasinghe met regularly to discuss progress, and soon observed the benefits, as LECO eventually satisfied all expectations. It was one of the outstanding successes of the Integrated National Energy Plan that Munasinghe presented to the President in 1984.

Highly skilled LECO staff gearing for routine maintenance activities (LECO file photo).

Teething Problems of the New Distribution Utility

Taking Over Assets and Service Areas

LECO had a nucleus of enthusiastic corporate and branch staff, but it did not possess the managerial or organizational resources necessary to service about 50,000 customers it planned to absorb in the first phase of its development. To rehabilitate and expand the rundown and overloaded council networks, the company needed to quickly set up and operate an extremely dedicated and efficient organization.

Existing council electricity networks needed a complete rehabilitation to improve the quality of supply, reduce losses, and improve the level of service to customers which required considerable investments. LECO did not have sufficient funding either for the rehabilitation or expansion of networks. Concurrently, there was an urgent need to improve the financial management of LECO and review the transfer price of electricity from the bulk supplier CEB. In addition, the interim accounting and billing systems needed to be quickly replaced by systems capable of handling rapid increase of customer volumes. All of these required considerable investments.

Capital, human resources, organization, accounting, and network development were all required at once.

Within a few months after setting up a skeleton staff, LECO took over its first service area in 1984, the Kotte Urban Council, a network serving 13,000 customers which was incurring 30% energy loss. The Government of Sri Lanka was developing Kotte as the new administrative capital of Sri Lanka. The new parliament building was already operational

Service area of Project I. The first councils LECO took over and rehabilitated (LECO file photo).

in Kotte and more administrative buildings were under construction.

LECO planned to take over another council network a year later in 1985. Commencing from the councils adjoining Kotte, LECO progressed by taking over council networks according to LECO's proposed service areas.

This was the first stage of taking over councils, and it was planned that the rebuilding project of networks would be completed by 1987. LECO took over 10 council networks by 1987.

The government, in late 1983 while LECO was being formed, requested technical assistance (TA) from ADB to support establishment and operation of LECO. Less than 1 year after LECO was incorporated, ADB prepared the first project and in January 1985, ADB approved the first loan of $12.4 million (later increased to $14.8 million) to finance Project I, "Secondary Towns Power Distribution Project," while the balance of funds for the $30.0 million project was provided by LECO and the government.[3] LECO estimated it would serve 54,000 customers, both existing and new, when Project I was completed across 10 council networks taken over during the first stage: Kotte, Kotte–Mount Lavinia, Kotikawatta, Maharagama, Peliyagoda, Ja-Ela town, Ja-Ela suburbs, Mahara, Dalugama, and Welisara.

> ADB provided technical assistance and financed the three-stage project of rebuilding the medium-voltage and low-voltage network.

The project constructed 33 kV and 11 kV lines, constructed low-voltage lines, replaced distribution transformers, rehabilitated customer connections, and replaced electricity meters.

The new distribution utility was not without problems. Successful bidders to the project did not honor the bid prices, requiring repeat bidding; material suppliers were lagging behind the delivery schedules, requiring cancellation and resorting to international bidding; poor performance of contractors required reallocating work to other contractors; and the political and security situation in the country was not conducive for outdoor work.

The project was completed in 1990, after a 2-year delay, but achieved its intended objectives. By the end of 1989, LECO energy losses were down to 16%. Customers served had risen to 179,000.[4]

While the successful implementation of Project I was in progress, the second project was proposed to be implemented from 1987 to assist the continued expansion and growth of LECO.[5]

Pole-mounted transformer. In 1985, LECO pioneered fixing distribution transformers on a single pole (Photo by Thisara Karunathilake).

---

3   ADB. 1984. *Report and Recommendation of the President to the Board of Directors: Proposed Loan to the Government of Sri Lanka for the Secondary Towns Power Distribution Project.* Manila.
4   ADB. 1990. *Completion Report: Secondary Towns Power Distribution Project in Sri Lanka.* Manila.
5   ADB. 1987. LN 870-SRI: Secondary Towns Power Distribution Project II.

New poles, pole-mounted maintenance-free transformers, and new cables were common sights in LECO service areas. Sectionalizers and auto-reclosers were fixed in the rebuilt medium-voltage network for improved protection and higher reliability. Reliability was increased; losses were reduced. New customers were connected.

The total cost of Project II was estimated as $52.2 million, of which $35.1 million was financed by ADB. Although Project II was expected to be completed by June 1991, the completion was substantially delayed by 3 years.

Project II increased the distribution voltage from the dismal 85 V (worst-reported at that time) in many areas to its design level of 230 V, reduced outages by two-thirds, and reduced system losses from about 30% to less than 10% in the council networks taken over. Institutional strengthening led to improved billing and collection systems and faster maintenance services.

Networks of 22 more councils were taken over during 1987–1990.

With the implementation of Project II, financial losses of LECO at its inception gradually reversed. Project II served as a powerful example of the viability that can be achieved with enhanced autonomy to a utility with dedicated staff, and with a strong and independent decision-making structure, despite being a government agency.

Certain network augmentations were not included in Project I or Project II, considering the funding limitations. By 1995, growth in customer demand in certain locations surpassed the originally planned capacities. Additionally, LECO was expected to provide electricity to areas which were not previously electrified, conforming to the government's plan of reaching the national target of 50% electrification by 1995.

The final project financed by ADB commenced in 1995, Project III.[6] The project included rural electrification, expansion of the upstream 33 kV distribution system, and expansion of the LECO distribution system. The total project cost was estimated at $115.3 million, of which ADB financed $71.2 million.

Mapping before Taking Over

Most councils did not have maps showing their electricity distribution system. Even if maps were available, they were outdated. LECO survey teams sketched the distribution lines and captured the details during the takeover.

LECO developed 1:5,000 maps using 1:50,000 maps purchased from the government survey department. LECO further developed 1:2,000 hard-copy maps to include all the system data and record customer data. Digital maps were not available in the 1980s.

All networks were mapped manually in the 1980s before these were digitized.

---

6   Asian Development Bank. 1997. LN 1414-SRI: Secondary Power System Expansion (Sector) Project.

LECO ran a well-equipped drawing office to develop base maps and then to mark surveyed data for further planning. Planning engineers studied the survey data and prepared requirements for the upcoming takeover of each network. The transfer process, therefore, was technically smooth. The subsequent operation, maintenance, billing, and revenue collection by the respective branch offices commenced soon after the takeover.

During the rehabilitation of networks, the drawing office developed maps and proposed drawings for all the new developments and finally the "as built" drawings were developed.

Hand-drawn maps were then passed on to the project design team for designing the new low-voltage distribution system.

Asset Valuation

After the maps of the existing distribution system were completed with the available data, two planning engineers physically verified the assets with the council staff. Each distribution feeder was physically verified for all the components, including transformers serving both retail and bulk customers, switches, switching points, and distribution lines. Many irregularities and ambiguities were found when the surveyed data were matched with the information provided by the councils. Usually, these were solved through an additional field visit.

Shares of LECO were issued to each council, in lieu of the assets taken over. If councils owed any outstanding payments to CEB by way of unsettled

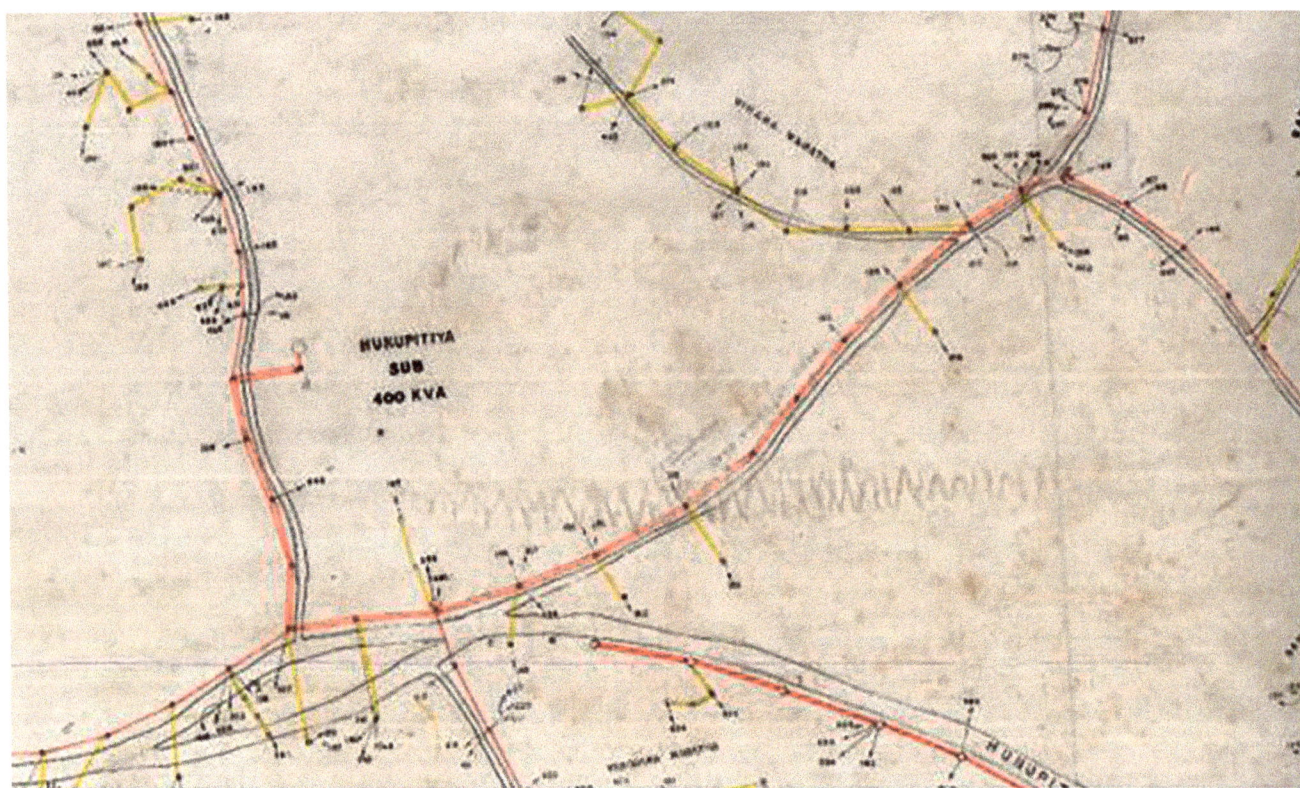

Hand-drawn maps documented all distribution lines and customer locations (LECO file photo).

> "I managed the difficult discussions with councils. I remember visiting all the councils, and meeting with councilors and chairmen. I negotiated many aspects: network extent, transition arrangements, staffing, and asset valuation. The asset valuations by the government valuer in accordance with the regulations, was below the expectation of each council. Extensive discussions and negotiations ensued reaching agreement at the end."
>
> Shanthi Amaratunga,
> chief negotiator of LECO,
> a chartered engineer

electricity bills or otherwise, an equivalent number of shares commensurate with arrears were issued to CEB. Objections to the takeover of council networks were settled through discussions and, in the case of one council, through a court decision.

After the takeover of all the identified council networks by 1995, CEB owned 54% of the total shares of LECO (inclusive of initial shareholding and shares allocated in lieu of bills issued to councils), while the government treasury owned 43% (in lieu of duty concessions for imports under the first two projects), and UDA owned the remaining 3%.

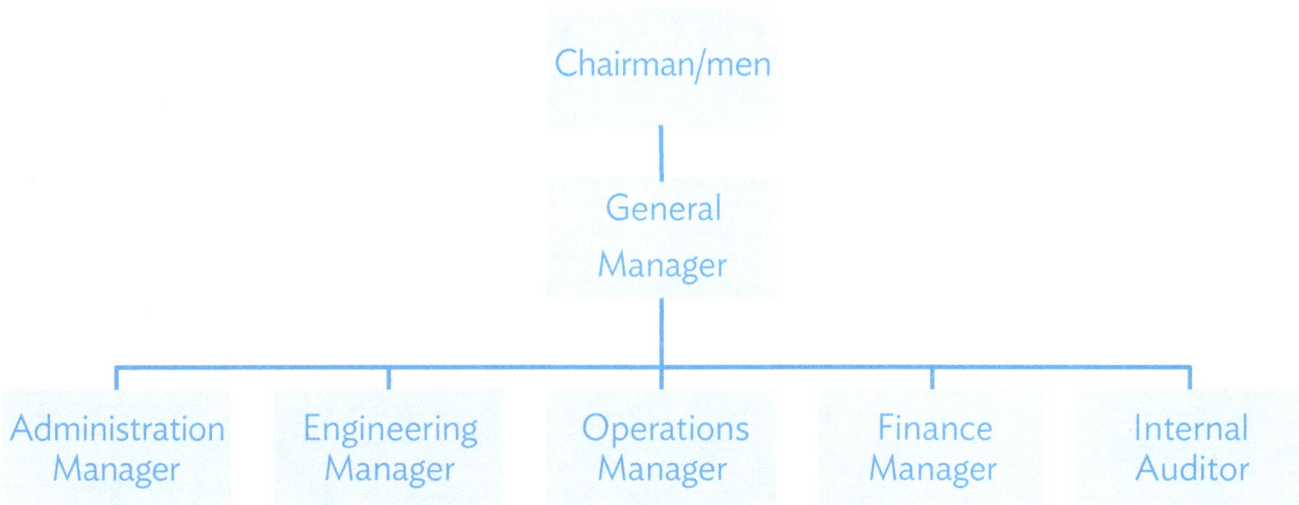

```
                    Chairman/men
                         |
                      General
                      Manager
                         |
  ┌───────────┬──────────┼──────────┬───────────┐
Administration Engineering Operations  Finance   Internal
  Manager      Manager     Manager    Manager    Auditor
```

LECO organization chart (Source: LECO company profile, short and long-term plans, 1994).

Innovation. LECO's innovativeness in utilizing the limited space in urban areas (LECO file photo).

LECO: A Pioneering Mission in Electricity Distribution

H. S. Subasinghe, chartered electrical engineer, rose through the ranks to become the general manager, CEB, the national power utility of Sri Lanka. He retired from CEB in 1983 and was appointed chairman/men of LECO in September, a position he held until 2001. He functioned as both chairman/men and chief executive officer (CEO) over 1983–1990. At the time of this interview, he was 95. He passed away in February 2021 at 97.

W. A. L. W. Amarasiri Perera, chartered electrical engineer, joined LECO in 1986, after a career in CEB in Sri Lanka and in Africa. He initially worked on Projects I and II to take over and rehabilitate council-owned networks. In accordance with articles of association of LECO and agreements with the Asian Development Bank (ADB) to separate functions of chairman/men and CEO, Perera was appointed as the general manager of the company in January 1990. He held the position for 17 years up to February 2007, the most crucial development phase of LECO. Perera joined the discussion with Subasinghe.

As the founding chairman/men of LECO, a position he held for 17 years in 1983–2001, steering LECO during its formative years and perhaps the most difficult period, Subasinghe recalled his unique experience.

In the first round, 10 council networks, including Kotte, the emerging new administrative capital of Sri Lanka at the time, were taken over by 1986. There was no resistance from the 10 councils, and they agreed to hand over their assets and become shareholders of LECO. This may have been because the President himself was the minister-in-charge of the subject of Power and Energy. However, during the second round, the councils did not want to become shareholders but were paid for their assets.

Naturally, during this takeover process, the asset valuation was a difficult process which needed extensive discussions and negotiations. The asset valuation carried out by the government valuer in accordance with the regulations was below the expectation of the councils. However, the chief negotiator of LECO, Shanthi Amaratunga, a chartered engineer, visited each of these councils and successfully managed the discussions, eventually leading to agreements. These networks were then taken over and later rehabilitated using funds allocated under two ADB-financed projects.

The LECO expansion stopped after taking over 22 councils, even though it was meant to take over more, with 219 councils spread all over the country at the time. This abrupt stop in LECO expansion was mainly because of resistance from CEB. Thus, LECO's expansion stalled beyond the Project II area. By the time phase III of the project (Project III) commenced, LECO had totally replaced the old distribution systems, with new 11 kV and low-voltage systems with LECO's own assets.

Around this time, another aspect LECO investigated was the optimum distribution loss levels in the LECO-served areas. Contrary to the belief at the time that the optimum distribution losses were about 10%, a study conducted under the leadership of the system development manager on the direction of the general manager LECO, concluded that losses can be reduced to 6% based on techno-economic evaluation. LECO moved ahead with the relevant investments to achieve this target. This initiative led to network losses being reduced to 8% by 1993, and further reduced to 6% by 2002. By 2019, LECO had achieved a loss level of 4%.

Subasinghe recalled, as the founder Chairman/men, the President's advice to him was to run LECO as a private enterprise. The senior staff of LECO at its initial stage were all ex-employees of CEB, but many with international utility experience. They were persistently working harder to achieve their goals. LECO was especially careful in hiring staff, to handpick the staff to run the company efficiently. LECO continued to hire staff during times when there was a moratorium on hiring for mainstream state-owned enterprises, and successfully defended such actions. At the same time, LECO never poached on CEB staff, but on the other hand, it allowed its own staff, mainly engineers, to join CEB if they so desired.

Perera spoke on the issues LECO faced initially to rehabilitate or replace the distribution network, which needed long periods of power interruptions. The dilapidated distribution network had to be replaced with new lines and transformers. Obviously, customers were unhappy over prolonged power outages, and complaints moved right up to the political hierarchy. At least on two occasions, LECO's actions to interrupt electricity supply during the rehabilitation process were criticized in Parliament. On one occasion, LECO was criticized for disconnecting supplies from those using electricity illegally to serve festive illuminations. On the second occasion, the complaint was more general, that it was unnecessary for LECO to pull down the network and build a new one because Sri Lanka was poor and underdeveloped and, hence, does not need such "gold-plated" investments on networks, the protestors said. However, as service improvements were achieved due to rehabilitation with the correct strategy, such complaints gradually turned to compliments on LECO's performance.

Perera continued on the complexities of taking over council networks in full. In some cases, LECO had difficulties in making its network geographically contiguous, because some of the geographical areas under a particular council had been already managed by CEB due to inability of the council to serve those areas. Therefore, the question was if the council network was transferred to LECO, should the CEB-managed section within the council territory be handed over to LECO too? Most often that did not happen, and even today, there are a few councils in which certain segments are served by LECO and others by CEB. This cohabitation, though undesired, has not caused any problems, other than perhaps 33 kV (CEB) and 11 kV (LECO) lines laid in the same area which could have been avoided if a single distributor had the total control. Subasinghe added that the two institutions agreed to exchange their networks in two towns for reasons of technical convenience, and to improve network reliability through better connectivity. LECO fulfilled the agreement, but CEB did not.

Perera recalled, in 1996, the government plan to sell a majority stake of LECO to a foreign company, an initiative of the then chairman/men of CEB. The idea came up through him to sell LECO to a foreign company, reportedly for a sum of SLRs3 billion (about $55 million at the time). However, LECO management expressed its desire to sell LECO shares to the public, though many did not agree with that idea either. Amid all this, CEB engineers went on strike, protesting against "selling off LECO." The strike caused a national blackout that lasted for nearly 2 days. With the strike, the idea of "selling off LECO" disappeared into just a memory.

With electricity reforms high on the agenda in 2002, there were various proposals to restructure the entire electricity industry. The aim was to have independent distribution units to manage the industry, and five distribution units were proposed to be established to implement the reforms. LECO was to be either one unit or be shared across the five units. The plan was to form a company for each unit. In other words, the entire distribution system in Sri Lanka was to be restructured into five government-owned companies similar to LECO. CEB agreed and proceeded with an internal reorganization to match with the planned restructuring, by effectively creating four distribution units, with all the infrastructure and staff, creating many new managerial positions. The restructuring process was halted in 2004, until limited restructuring was legally established in 2010 with the new Electricity Act 2009. To date, LECO's service area remains what it was at the commencement of reforms.

When Perera was appointed the general manager in 1990, the rehabilitation projects were in full swing and they needed equity and project management staff, and LECO was experiencing financial difficulties. LECO had an outstanding balance of many millions of rupees to be paid to CEB, who provided supply in bulk to LECO. Those were very hard times for LECO, but within 3 months into his tenure, as a priority, LECO settled all the arrears to the bulk supplier, CEB.

Perera recollected how LECO carefully managed to bring down its costs, while continuing to improve customer service. In the early days, when a customer complained, LECO's response was to send a technician on a motorcycle with a sidecar that carried a helper and tools. This was naturally a faster response which customers appreciated. If the problem could not be solved at that level, then a full maintenance team was called. On the other hand, CEB's practice was to send a truckload of technicians, taking a longer time to attend to the complaints. LECO's arrangement of technician-on-motorcycle was partly because LECO could not afford to buy many vehicles.

The pricing of electricity to LECO had always been a debatable issue. It was always a negotiated tariff and negotiations were painful for everybody. On one occasion, CEB increased the agreed tariffs, without even notifying LECO, leading to around SLRs500 million of additional cost annually. Only the secretary, Ministry of Power and Energy had a scientific approach to price calculation. Many objected with various arguments. Perera requested his senior staff to prepare a professional calculation methodology for tariffs, and he even presented a paper at the Institution of Electrical Engineers on "Electricity pricing to infuse competition." LECO then made a tariff submission based on the methodology given in the paper.

Calculations by LECO and by the Secretary, Ministry of Power and Energy were ignored. CEB placed its own calculations and proceeded with approval despite resistance from LECO. There were many tariff methodologies, studies, and calculations over the years. Perera thought the tariff methodology currently in practice, adopted as a result of the Electricity Act 2009, was by far the best, giving a predefined return to LECO on its assets, to earn adequately to finance the implementation of the approved investment plan.

Both Subasinghe and Perera spelled out their vision for LECO. They were unanimous in their resolve that LECO needs to venture into new technology in all spheres of its activities: metering, demand management, and distributed generation; and continue to improve its services to customers at ground level. LECO should continue to maintain its higher levels of network reliability and lower levels of losses, setting an example to other utilities in Sri Lanka and Asia. Most importantly, the shareholders of LECO (they are all government agencies) need to maintain an arms-length relationship, to enable LECO to continue with "private sector efficiency" backed by the stability afforded by government ownership. This requires the appointment of independent and efficient professionals of high integrity to board positions.

"Continue to hire handpicked efficient staff to develop and manage engineering and commercial functions of LECO," Perera concluded.

Lanka Electricity Company was a Successful Experiment

G. A. D. Sirimal's thoughts as Secretary assisting the Committee on Electricity Distribution Reforms

G. A. D. Sirimal was the assistant secretary of the Ministry of Power and Energy for over 2 decades from 1980. He assisted the committee that studied the options to resolve the problems faced by CEB in relation to local government institutions such as city councils. Subsequently, he was closely associated with the formation of LECO, and now in his retirement, continues with his mission to enlighten the public on the importance of electricity industry reforms and professional management.

The local authority electricity distribution businesses were bleeding because of poor overall operational and financial management. Dues to CEB, the bulk power supplier, had been mounting for years. The government at the time, with a strong mandate for rapid economic development, needed to examine options. One obvious option was to ask CEB to take over distribution operations in councils. It would have been an uphill task because with such a takeover, the customer base of CEB would have doubled to half a million within a short period. CEB at that time was not equipped to accommodate such a massive task.

LECO was an experiment at the time. One main objective was to see how the concept of a government-owned company worked, so that in other provinces, similar companies may be set up later. Finally, LECO limited itself to densely populated coastal districts of the western province and the southern province, which were areas with a high customer density. As electricity costing and pricing were not streamlined as they are now, LECO surely found a comfortable space only in the zones with high population density.

From the beginning, LECO management was determined to operate it as an independent company. The idea in forming this company owned by three shareholders, i.e., CEB, Treasury, and UDA, was for LECO to be distanced from political interventions. LECO was only answerable to its shareholders who, in turn, were government-owned. The strategy worked. Along with the formation, LECO managed to build a strong team of engineers and managers. Initial staff was carefully selected from among handpicked retirees from CEB and those with international experience. Council staff was only selectively hired based on their qualifications and experience. This strong team brought a new life to the distribution sector with high levels of customer service, and with prompt attention to customer queries and breakdowns. Similarly, the articles of association allowed LECO to go into other businesses in later years, once its mission to deliver a high quality of electricity supply and services was achieved. A meter manufacturing facility is already operational.

"I am extremely happy that LECO came into being, and for the opportunity I got to contribute to its formation. LECO's high level of performance through its 4 decades of existence is a testimony that LECO, the experiment, was indeed a success," concluded Sirimal.

A New Approach to Engineering and Customer Care

Administrative Independence

LECO was not established to be another "board," a typical structure of government-owned utilities that prevailed in the 1980s. LECO was a company, independent of government controls and is run free of restrictive circular administrative instructions that retard innovation.

Administrative independence throughout the first 20 years since its incorporation enabled the company to modernize its operation and structure. LECO was free of government controls on staff recruitment, salary and benefits, procurement procedures, and all other restrictive activities.

> LECO instituted a culture toward improving customer service and relationships.

Many innovative procedures were already introduced by the mid-1980s:

(i)   instant billing,

(ii)   providing new electricity supply connections without deposits against future bills,

(iii) a wide choice of places to pay the bill,

(iv)   customer consultative meetings in branches,

(v)   customer complaints resolved at branch level,

(vi)   performance assurances to customers,

(vii) a central distribution control center to handle both customer complaints and operations control, and

(viii) producing monthly service reliability reports.

LECO control center in action. Attending to customer complaints (Photo by Tharindu de Silva, LECO).

LECO's low-voltage design philosophy moved away from the traditional thinking of centralized transformers and long low-voltage lines. LECO focused on distributed, smaller-capacity transformers and shorter low-voltage lines. The traditional two-pole structure for transformer mounting would be more expensive and occupies more space. Therefore, LECO designed a concrete pole to mount the transformer. This resulted in a faster approach with a slight modification to the network to meet requirements of any new customer, retail or bulk, by placing additional transformers atop already installed poles. If the pole strength is not sufficient, a replacement of just one pole was needed to place the transformer. No additional land would be required, unlike in the conventional utility practice. LECO mounts 100 kilovolt-ampere (kVA), 160 kVA, or 250 kVA transformers on a single pole.

LECO deviated from the practice in regional utilities. Transformer mounted on a single pole saves urban road space since 1985; poles do not need a strut or a stay wire to support, saving precious garden space (Photo by Thisara Karunathilake).

All low-voltage lines of LECO were insulated right from the inception in 1985. Insulated medium-voltage lines (the upper conductor in this photo) are used in congested urban areas to improve safety and reliability (Photo by Thisara Karunathilake).

Customer service centers catering to the needs of customers (Photo by M. A. Pushpa Kumara).

Distribution Planning: Key to a Successful Utility

At a time when many distribution utilities in the region did not consider it important to plan and optimize power supply networks at medium and lower voltages, LECO established a planning department and planning tools, including software, to conduct load flow studies and voltage profiling for its network, right up to the service connection point of each customer.

Planning Criteria

- Voltage regulation:
  11 kV: 4%
  400 V: 2%.
- Reliability criteria:
  11 kV network satisfied the N-1 reliability criterion for metropolitan and urban areas
- Allowable energy loss:
  11 kV: 1%

Target total LECO network loss: 4.02% (2020).

LECO realized that to be in the forefront of the distribution business, productivity should be improved and the best way to achieve this is to computerize as many processes as possible. Staff produced in-house software for many analytical tasks including transformer load management, load flow analysis at low voltage, material scheduling, pole selection, demand forecasting, and financial and/or economic analysis of new electricity distribution schemes.

LECO developed its own geographic information system (GIS) software in the 1990s and introduced procedures to use global positioning system (GPS) technology to collect and transfer data to the GIS. The entire network has been mapped on GIS since the 1990s. Network studies precede the connection of each new customer. LECO used this own GIS and the data capturing methodology for nearly 20 years

until an industry-standard software was adopted in 2017.

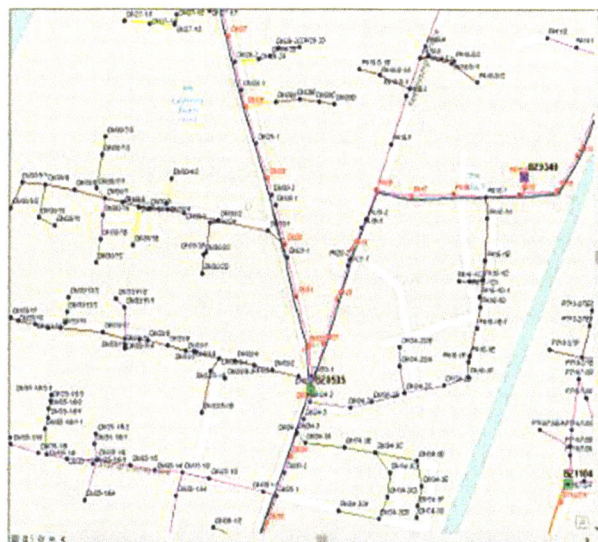

Use of NEPLAN 360. Fully integrated GIS-enabled software is used today for network planning and customer support (photo is an extract from LECO's GIS database).

Predicting network constraints amid the growing demand was part of the routine planning work of LECO. Solutions introduced to network constraints are many:

(i) introducing new load-break switches,

(ii) upgrading existing lines,

(iii) introducing new distribution lines or express lines,

(iv) conversion of 11 kV lines into 33 kV feeders,

(v) identifying potential new 33 kV injection points, and

(vi) identifying new grid substations to be included in the long-term transmission and financing plans.

By the year 2000, 17 years into its existence and with a customer base of 330,000, LECO reduced distribution network losses to 8.5%, when the parallel utility CEB was saddled with losses estimated at 15% in distribution.

While LECO moved toward the standards of a world-class distribution utility, the physical planning of the country did not move to the expected standards in urban planning and space management.

LECO currently plans to establish a 33 kV backbone network to connect existing primary substations and operate a combined 33 kV and 11 kV mixed system. Transfer of large customers to the 33 kV network, converting selected overhead lines to underground cables, and introducing network automation are all being planned.

> Meticulous planning by design teams has enabled the company to improve the quality of supply. Planning enables engineers to foresee and initiate action to solve emerging network constraints in this fast-growing urban utility.

LECO has been able to maintain good voltage regulation due to its farsighted policy of adhering to the concept of "smaller transformers–shorter lines" and following the guidelines laid down in the design of low-voltage distribution schemes.

LECO has been monitoring its system reliability in compliance with the performance standards laid down by the government under the Electricity (Distribution) Performance Standards Regulations of 2016.

Good monitoring of reliability data enables LECO to identify trends. LECO is currently observing an increasing trend of outage events and duration of outages.

LECO is expected to provide its customers with reliability levels on par with the capital city of Colombo. The city has fully underground transmission and distribution networks, and LECO serves urban areas to the north and east of Colombo.

> Suburban communities expect a quality of supply comparable with the capital city.

Early identification of a rising trend in outages allowed LECO to formulate a project to strengthen the eastern suburbs with an underground network.

LECO central distribution control center (Photo by Tharindu De Silva).

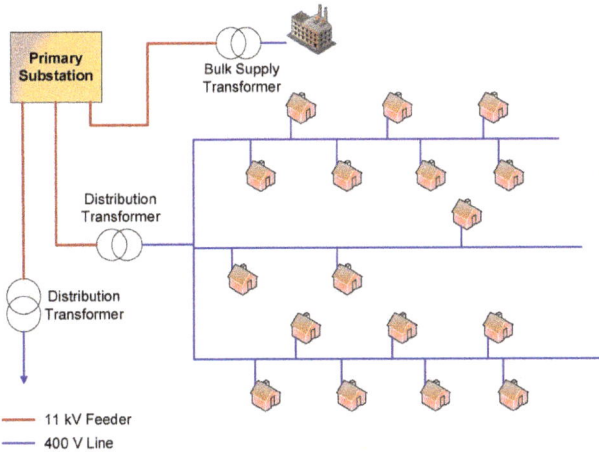

11 kV Feeder
400 V Line

Larger Transformers, Longer Distribution Lines; that was the policy up to 1980s (Sketched by Muditha Karunathilake).

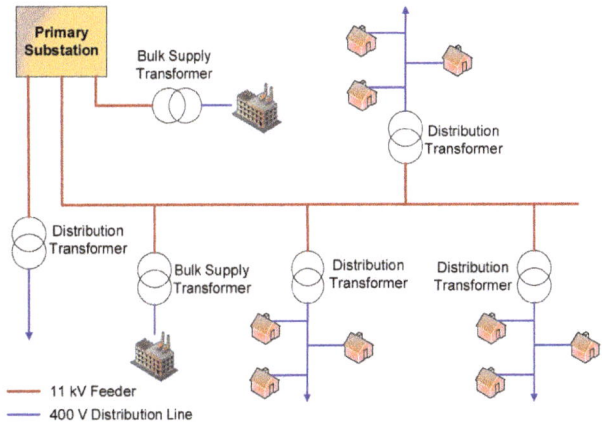

11 kV Feeder
400 V Distribution Line

Longer 11 kV lines supplying distribution transformers at no more than 1 km intervals, and shorter distribution lines, enabled LECO to minimize energy losses and to provide a quality supply (Sketched by Muditha Karunathilake).

Reliability indexes are calculated using a database management software. For every outage, the number of customers interrupted is known. This information is acquired in advance and saved in the database. Outage data collected is used along with the number of customers interrupted to calculate reliability indexes. The indexes system average interruption duration index (SAIDI), system average interruption frequency index (SAIFI), customer average interruption frequency index (CAIFI), and customer average interruption duration index (CAIDI) are calculated and monitored.

LECO today develops 10-year long-term development plans and continues the legacy of planning as a key to success.

"Planning, using state-of-the-art techniques at the time, and their implementation on the ground enabled LECO to quickly reach its targets in engineering performance and customer satisfaction."

Jayasiri Karunanayake,
System Development Manager
LECO, 1988–2000.

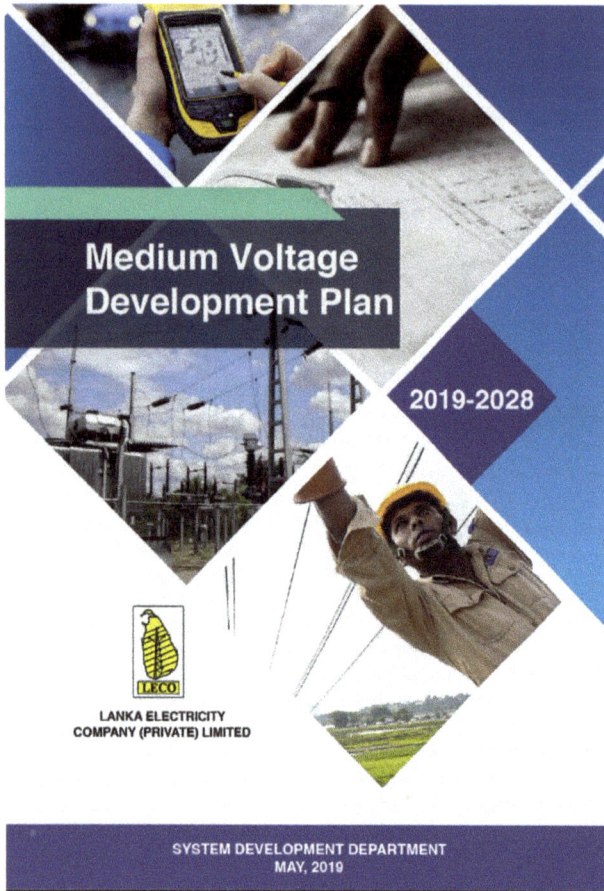

Medium Voltage Development Plan

2019-2028

LECO

LANKA ELECTRICITY COMPANY (PRIVATE) LIMITED

SYSTEM DEVELOPMENT DEPARTMENT
MAY, 2019

Cover page of a recent medium voltage development plan, 2019–2028. The plan is updated once in 2 years.

Innovations in line designs save urban space and make the lines safer (Photo by Thisara Karunathilake).

Treasures from the Archives

Smaller Transformers, Shorter Lines

Planning veterans L. C. Amaratunge and Frank Perera analyzed the technical, financial, and economic benefits of "large transformers and long, low-voltage feeders" versus "small transformers with short, low-voltage feeders," in their paper published in 1993. The study confirmed the LECO practice was correct both technically and financially and provided the vital basis to forge ahead with the concept in all LECO designs ever since.

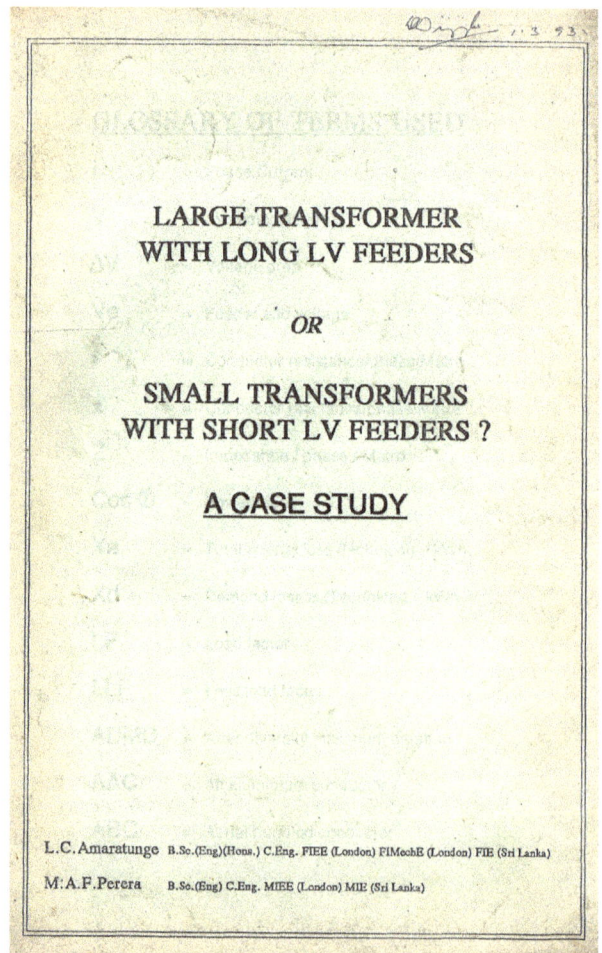

LARGE TRANSFORMER WITH LONG LV FEEDERS

OR

SMALL TRANSFORMERS WITH SHORT LV FEEDERS ?

A CASE STUDY

L.C.Amaratunge B.Sc.(Eng)(Hons.) C.Eng. FIEE (London) FIMechE (London) FIE (Sri Lanka)

M.A.F.Perera B.Sc.(Eng) C.Eng. MIEE (London) MIE (Sri Lanka)

A study in 1993 confirmed the LECO practice of using smaller transformers to be more economical (LECO file photo).

Protection Engineering Applied to Distribution

Recent Developments in Distribution System Protection and it's Impact on System Performance

J. Karunanayake   27th April 1999
Lanka Electricity Company (Pvt) Ltd.,

1.0   Introduction

Advances in protection relaying during the last two decades can only be compared to what took place in the very early days of the industry. In that era, a power system usually consisted of a small generator supplying a small local load. Even these had its own protection with the first known scheme being a totally manual system which depended on one man, the attendant at the generating station. He had to watch the ammeters, feel the conductors and had to open a great knife switch at the first sign of smoke.

Expansion of the power systems made the engineers to concentrate on researching for equipment to detect faults and disconnect faulty sections automatically. Protection relay was the result and this laid the foundation for the Power System Protection engineering which has become an indispensable arm in Powers System engineering.

2.0   Protection Relays

Earliest form of protection relays were electromechanical units, which operate on the power supplied to them through the current and voltage transformers. Next development was the static relay. These were mainly level detectors and timing circuits constructed with op amps, discrete components such as transistors, capacitors resistors etc.

In the next generation of relays microprocessors were used in place of level detectors, with some discrete components. Setting changes were effected through selectable switches and these relays introduced the facility of multiple characteristics, which is an extremely important milestone in protection engineering.

LECO routinely upgrades their Protection Practice (LECO file photo).

Power system protection specialist Jayasiri Karunanayake applied modern techniques to make the LECO network resilient and more reliable.

GIS Applications in the 1990s

LECO engineers, Jayasiri Karunanayake, K. K. Jayasundara, and S. D. C. Gunawardena, implemented GIS in the 1990s to map the LECO network.

Recent Activities in Space Science in Japan

By - Prof. Yasanori Matogawa
Professor, the Institute of Space and Astronautical Science
Director, Kagoshima Space Center

Since the first satellite "OHSUMI" in 1970, ISAS has launched scientific spacecraft at approximately once-a-year rate, independently of space application programs in Japan. This strategy of autonomy of science against application has contributed to be quite effective. ISAS successfully launched a space VLBI satellite HALCA (MUSES-B) by M-V launch vehicle on February 12, 1997. With the advent of this new launch vehicle, Japan's space science can be said to be entered into a new era, that is, a stage for more ambitious projects including lunar/planetary exploration. As a first batter, on July 4, 1998, ISAS launched the first Japanese Mars explorer NOZOMI (PLANET-B), which is

now on the way to Mars to arrive in October. Thus, as of the present stage, there are seven ISAS spacecraft alive. They are, in the order of their launch dates, SAKIGAKE: Halley's comet explorer, AKEBONO: aurora observation satellite, YOHKOH: solar observation satellite, GEOTAIL: satellite for geomagnetotail observation, ASCA: X-ray astronomy satellite, HALCA: space VLBI satellite, And NOZOMI: Mars explorer.

This report is a brief introduction about recent achievements and some near future programs in space science in Japan.

Application of GPS Equipment to Create a GIS for an Electricity Distribution Utility (LECO)

By – Mr. J. Karunanayake, Mr. K. K. Jayasundera, Mr. S. D. C. Gunawardena
Lanka Electricity Co. (Pvt.) Ltd.

Background: LECO started electricity distribution system mapping with a manual system. Base map information were from Survey Department maps. Due to various factors, system maps so created lacked the desired accuracy.

Mapping system was than computerized using AutoCAD. As digitizing with basic AutoCAD commands was laborious, AutoCAD was customized with Autolisp. Result was a user friendly package, enabling the combining the data-base and the mapping system, but this too had shortcomings.

GPS Solution: Use of GPS technology to collect and transfer data to a GIS was then

examined. Accordingly a pilot project employing a 12 channel GPS receiver was lunched.

Implementation: Creating the Data Dictionary is perhaps the most vital aspect in the data capture program. Identification of the data to be captured, which for an electricity distribution system are road names, electricity distribution lines, underground cables, transformers, switching equipment, poles, street lights, consumers etc. Each of these "features" of the distribution system has many attributes.; eg. Distribution line feature can have attributes such as feeder names, conductor sizes, voltage levels etc.

LECO initiated GIS in Distribution (LECO file photo).

Doorstep billing was practiced since 1984 (Photo by Duranka Menerigama).

Electricity Pricing

Electricity pricing was also a subject of intense discussion in 1993, before modern regulatory practices were implemented decades later.

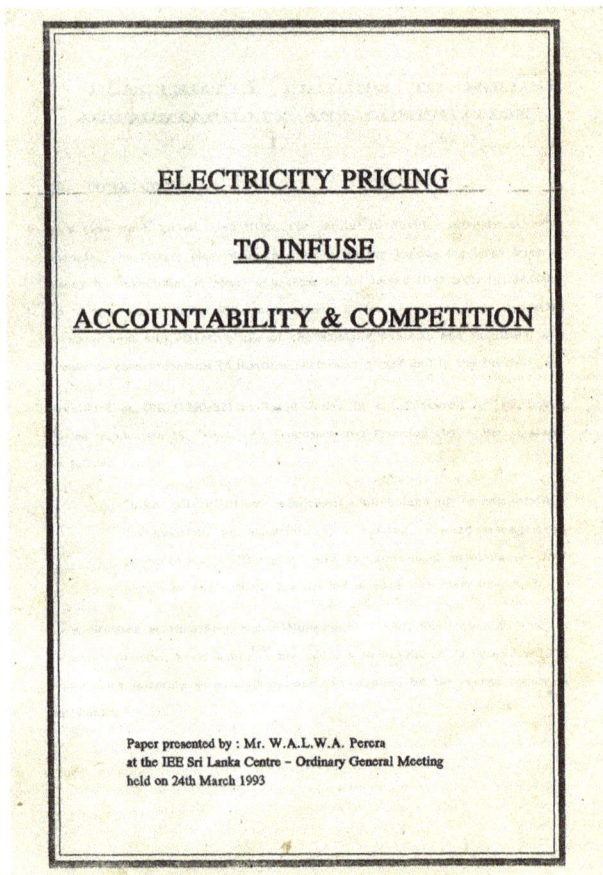

ELECTRICITY PRICING

TO INFUSE

ACCOUNTABILITY & COMPETITION

Paper presented by : Mr. W.A.L.W.A. Perera
at the IEE Sri Lanka Centre – Ordinary General Meeting
held on 24th March 1993

Electricity pricing study of 1993 (LECO file photo).

Improving Customer Service: An Example to Others

A. D. L. Gunawardena is currently the customer services superintendent in Maharagama, a fast-growing suburb of Colombo. He joined LECO in 1984 and served in many positions in Kotte, Nugegoda, and Maharagama. He recalled the initial interventions LECO had in improving customer service and how it continued to lead, setting a benchmark for others to follow.

"Since our first customer was connected in 1983, we read customer electricity meters every month and immediately issued the bill at the customer's doorstep, a great intervention nobody could have imagined at the time," he said.

At the time of taking over council networks by LECO, customer services were not at acceptable levels. In a way, that helped building a strong foundation for LECO with improved customer service, based on a strictly customer-oriented policy in all ground-level activities. When customers experienced LECO's high-quality of service, they immensely appreciated it and were glad that LECO was established.

One important customer concern was how fast LECO processed applications for new power connections. LECO changed the previous practice: an application was reviewed within 3 days. Technical officers were assigned to provide cost estimates for the connection. For example, in Kotte, there were three technical officers to prepare estimates for new customer connections. After the estimation was completed, the customer was informed to pay. Once the payment was made, the connection was given within 10 working days. Since there were officers who were specifically assigned to prepare estimates and construction, LECO was able to provide a new connection within 14 days from the date of application. This was a remarkable achievement in the 1980s.

There were problems in the billing procedure of councils. They first read all the meters and later sent the bill to the customer by hand or by post. For the first time in the country, LECO initiated the policy of doorstep billing, by issuing the bill immediately after the meter was read. That concept was introduced by LECO, and it eliminated most of the problems in billing.

At the inception of LECO, complaints were accepted centrally at the Kotte office and there was an officer to attend to them. Now, every customer service center accepts complaints. Any customer of any category can submit applications and complaints to the same office. Customers had to visit a branch office only to make a payment, but that too was soon simplified in the late 1980s by accepting payments at a wide range of outlets, including designated grocery shops.

LECO's policy of using insulated, bundled conductors for distribution lines and concrete poles of a superior design was criticized by politicians and the public alike to be "overdesigned." However, it was later proven to be a very good initiative of LECO. Damages and subsequent loss of electricity supply were drastically reduced when bare conductors used by councils were replaced with bundled conductors, fixed on newly designed poles. The pole, specially designed for LECO, could be fixed without a stay-wire or a strut which would require land space in the vicinity. This was greatly appreciated by landowners, particularly those with limited garden space in congested cities.

On attending to breakdown of supplies in the early days until 1985, since there was no radio communication, complaints were written on paper and handed over to the maintenance team. There were three people assigned to a breakdown gang. In some cases, a technician was also the driver who had been given permission to drive. If they could not handle the situation, another team with heavy equipment would be called to take over. With radio communication and computerization, the manual record was followed by its being entered into the outage management system and communicated to the maintenance crew over the radio.

Interestingly, there was no formal contract for electricity supply between the customer and the council until LECO introduced one. That new agreement was the only thing some customers protested on.

Self-service payment kiosks are now available at many LECO offices. LECO staff guide and encourage customers to use the facility (Photo by Duranka Menerigama).

As in the case of most new institutions at the start, there were many challenges. LECO did not have proper vehicles to transport poles at that time. The poles were carried on rented bullock carts. After the takeover process, several distribution lines were vandalized, possibly by persons who did not like the LECO takeover. LECO gradually addressed such shortcomings and obstacles and emerged stronger to serve the customers.

One of the major design changes LECO adopted was a feeding arrangement with longer 11 kV lines and shorter 400 volts (V) lines. The low-voltage feeder was made shorter than 0.5 kilometers, with 50 kVA or 100 kVA distribution transformers placed at frequent intervals, to step down the electricity supply from 11 kV to 400 V. The concepts of using bundled conductors and these new smaller distribution transformers mounted on a single pole were brought in with LECO management's full acceptance. These design changes drastically reduced excessive voltage drops, and both technical and commercial losses. LECO's practice adopted in 1980s, which was a major exception at the time, has now become the norm all over Sri Lanka and indeed, in the rest of Asia.

Planning a Distribution Network for Many Years to Come

K. K. P. D. Jayasena has been the chief draftsperson of LECO for the past 10 years. She joined LECO in 1984 as a trainee drafter at head office. After being confirmed in her appointment to a permanent position in the company, she served for 4 years at Kelaniya, during the formative years of LECO. She explained how the drafting team assisted the network to be planned and drawn, and subsequent digitization.

It is always interesting to go down memory lane and reflect on how the staff really started their respective work in LECO. Draftspersons developed all drawings manually in the early days. Later, they shifted to AutoCAD customized using AutoLISP and went on to develop geographic information system software, while introducing data collection using global positioning system (GPS) devices. At the inception, there were no maps in the government's survey department which could be used for this purpose. Draftspersons prepared their own maps using topographical sheets. Then the Kelaniya branch was computerized in 1986, as the first step to computerize the maps.

In this exercise, Pidurutalagala, the highest point in Sri Lanka, was taken as the reference and all customer numbers were defined based on this reference. The geographical area was entered into a grid and each customer's number was based on the grid number. LECO staff then visited each customer to collect the meter and account numbers, and plotted the network manually within the grid. In this manner, draftspersons were able to build the database and the map which were then used to assign account numbers to new customers. With the availability of GPS in the 1990s, the coordinates were obtained from GPS and they proceeded with network digitization which was completed for the whole LECO system by 2003. From the mid-1980s, this database allowed identification of a customer with 95% accuracy, and the transformer and the low-voltage line serving the customer with 100% accuracy. With this system in place, customer service improved remarkably, particularly to shorten the time to attend to supply outages. Even with this system, not as sophisticated as the current system, which is far superior, LECO managed having excellent customer service, even in the 1980s.

Important maintenance. LECO staff maintaining a street light (Photo by M. A. Pushpakumara).

LOAD **200** kg

The Success Story in Electric Utility Management

How Was It Done?

LECO inherited the council networks with high technical and commercial losses. Losses may have exceeded 30%, but official records indicate 23% in 1984. LECO persistently worked toward reducing all types of losses: technical and commercial, including metering errors, accounting errors, and theft.

Whenever council networks were taken over, LECO overall losses increased. With well-planned network replacements supported with loans from ADB, LECO was able to progressively reduce distribution losses. LECO required 16 years to reduce distribution losses from 23% to 8%, and a further 20 years to reduce from 8% to 4% (see figure 2).

Key interventions for these achievements were installation of new meters to customers,

Figure 2: Energy Losses in the LECO Network

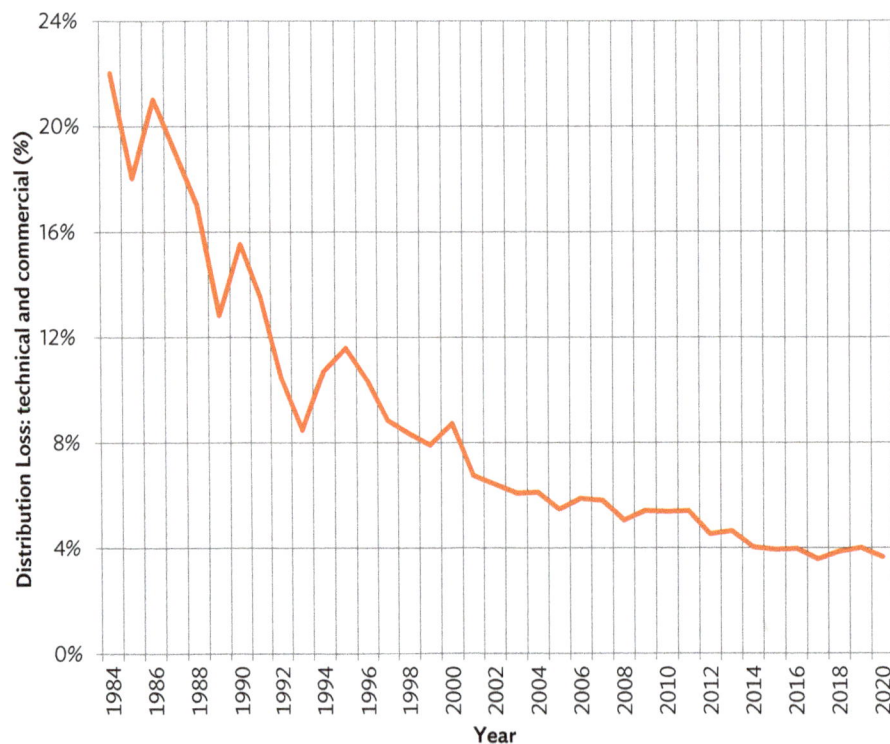

Source: Sri Lanka Energy Balance 2018, LECO Statistical Digest. 2020.

construction of new 11 kV feeders and distribution lines, installation of new distribution transformers, and fixing meters to all distribution transformers.

Furthermore, the use of aerial bundled conductors for low-voltage distribution eliminated electricity theft off the lines. Therefore, nontechnical losses of LECO by way of theft off the line is almost zero. A formal investigation process was introduced to handle theft. An investigation commences when information is received from any of the three specific sources: (i) technicians, (ii) revenue officers (meter readers), and (iii) anonymous sources. The detections are now reduced to less than 100 per year compared with several hundreds of offenses detected every year in the early days of LECO.

Meter tampering and theft are considered very serious matters, punishable by a court of law. LECO has successfully used the full strength of the law and its own internal procedures to ensure commercial losses are reduced to a minimum.

From a Loss-Making Start-Up to a Profit-Making Utility

Differences in retail prices that prevailed among council-managed areas and elsewhere in the country caused dissatisfaction among the electricity customers. Sri Lankans were seeking a uniform national tariff for electricity.

Maintenance works. LECO technical teams attending to routine maintenance work in a busy Colombo suburb (Photo by Duranka Menerigama).

LECO purchased electricity at the bulk supply tariffs determined by CEB, the bulk power utility. However, CEB was also the main distributor with a higher market share, currently nine times that of LECO.

The tariff for bulk sales to LECO was expressed in terms of retail energy sold with an allowance of 20% to account for energy losses in the LECO network at that time.

Network losses were reduced gradually owing to meticulous planning, management, and monitoring of the network. However, LECO was compelled to sell electricity below cost to low users and expected to recover such internal subsidies from high users.

It was impossible for LECO to generate surplus cash to meet its operational costs and debt repayment commitments. This was because the customer mix was biased toward lower blocks of consumption.

The company had to purchase electricity from CEB, but sell at prices lower than CEB's prices to its own customers. Interestingly, CEB itself determined the bulk supply tariff to LECO. Bulk supply tariff to LECO was determined after constant bargaining and disagreements with CEB. Without a proper electricity pricing methodology to allow cost recovery, LECO was suffering financial losses.

CEB wanted LECO to charge its customers at a higher retail tariff and continue to purchase at a higher bulk supply tariff. However, the government insisted that the same retail tariffs be maintained for both CEB and LECO customers, based on its policy of establishing a uniform national customer tariff across the country.

The Electricity Act No. 20 of 2009 provided the framework to establish a tariff methodology. With this introduction, all distributors in the country (CEB and LECO) receive electricity, in principle, at the same bulk supply tariff. Bulk supply tariff

comprises generation tariff, transmission tariff, and bulk supply and operations business tariff; and they are determined by the regulator, following the rules established in the tariff methodology.

ADB's technical assistance in 2010 enabled an electricity costing and pricing methodology to be established, ending years of debate on the cost of supplying electricity and the uncertainty over LECO's financial well-being.

LECO has demonstrated that distribution costs per kilowatt-hour (kWh) decline in real terms, as the utility sells more electricity over the same network. However, as the utility assets age, they must be replaced, and the network requires upgrades to meet the increasing customer expectations of higher reliability and to absorb distributed generation.

The tariff methodology followed by Sri Lanka since 2011 to determine prices for inter-licensee transfers was prepared with technical assistance from ADB.[7]

Since 2011, the transmission licensee and the five distribution licensees were allowed a 2% return on assets while depreciation, interest costs, operation and maintenance costs, and taxes are fully allowed. The methodology considered the disparities between the five distribution licensees (see figure 3) by way of customer mix and the extent of the network, and developed clear procedures to compensate each licensee. LECO is the distributor out of the five distributors, with the customer mix yielding the highest revenue per unit of electricity sold. The methodology, therefore, requires LECO to purchase at a higher "adjusted" bulk supply tariff.

---

7   ADB 2019. Sri Lanka: Clean Energy and Access Improvement Project; TA 7265-SRI: Capacity Development for Power Sector Regulation, Technical Assistance Completion Report, ADB, Manila. https://www.adb.org/projects/documents/sri-39419-013-pcr

Figure 3: Revenue Flow in Sri Lanka Electricity Industry

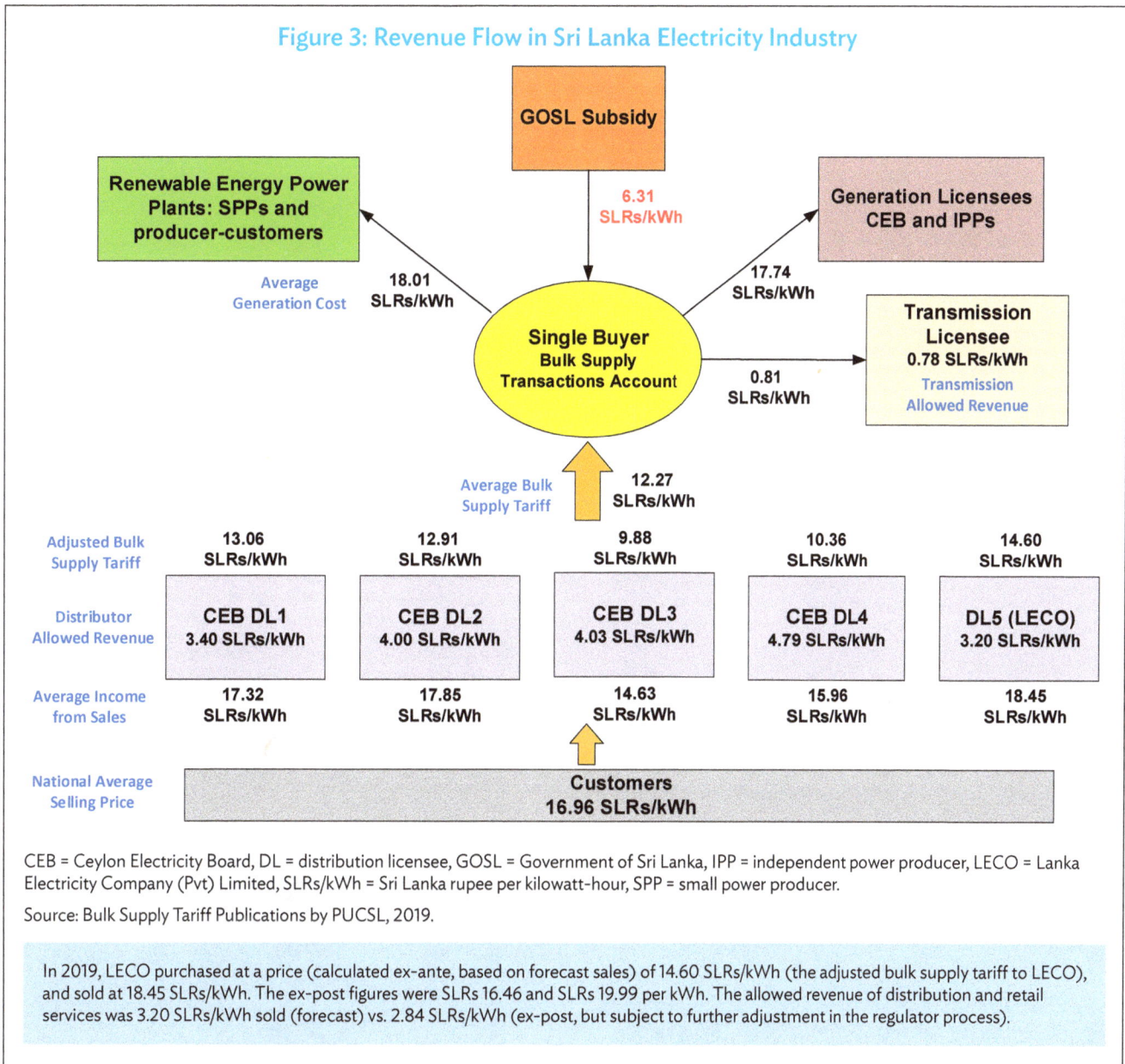

GOSL Subsidy

6.31 SLRs/kWh

Renewable Energy Power Plants: SPPs and producer-customers

Generation Licensees CEB and IPPs

Average Generation Cost — 18.01 SLRs/kWh

17.74 SLRs/kWh

Single Buyer Bulk Supply Transactions Account

Transmission Licensee 0.78 SLRs/kWh
Transmission Allowed Revenue

0.81 SLRs/kWh

Average Bulk Supply Tariff — 12.27 SLRs/kWh

	CEB DL1	CEB DL2	CEB DL3	CEB DL4	DL5 (LECO)
Adjusted Bulk Supply Tariff	13.06 SLRs/kWh	12.91 SLRs/kWh	9.88 SLRs/kWh	10.36 SLRs/kWh	14.60 SLRs/kWh
Distributor Allowed Revenue	3.40 SLRs/kWh	4.00 SLRs/kWh	4.03 SLRs/kWh	4.79 SLRs/kWh	3.20 SLRs/kWh
Average Income from Sales	17.32 SLRs/kWh	17.85 SLRs/kWh	14.63 SLRs/kWh	15.96 SLRs/kWh	18.45 SLRs/kWh

National Average Selling Price

Customers 16.96 SLRs/kWh

CEB = Ceylon Electricity Board, DL = distribution licensee, GOSL = Government of Sri Lanka, IPP = independent power producer, LECO = Lanka Electricity Company (Pvt) Limited, SLRs/kWh = Sri Lanka rupee per kilowatt-hour, SPP = small power producer.

Source: Bulk Supply Tariff Publications by PUCSL, 2019.

In 2019, LECO purchased at a price (calculated ex-ante, based on forecast sales) of 14.60 SLRs/kWh (the adjusted bulk supply tariff to LECO), and sold at 18.45 SLRs/kWh. The ex-post figures were SLRs 16.46 and SLRs 19.99 per kWh. The allowed revenue of distribution and retail services was 3.20 SLRs/kWh sold (forecast) vs. 2.84 SLRs/kWh (ex-post, but subject to further adjustment in the regulator process).

LECO is independent. The company has to manage all its operational expenses and depreciation; pay interest; and earn profits, with SLRs3.20 per kWh (estimated for 2019, in nominal terms) allowed to be retained out of the income from electricity sales.

In 2019, LECO purchased electricity from the grid at SLRs14.60 per kWh, the highest price among Sri Lanka's five distributors.

Table: Allowed Revenue for LECO's Distribution and Retail Supply

Parameter	In Constant January 2011 Prices					In Constant January 2016 Prices				
	2011	2012	2013	2014	2015	2016	2017	2018	2019	2020
Distribution revenue cap (SLRs million)	2,219	2,279	2,338	2,397	2,456	3,348	3,408	3,467	3,526	3,585
Retail services revenue cap (SLRs million)	296	303	325	332	324	487	498	508	518	527
Forecast Sales (GWh) for upfront tariff determination	1,198	1,241	1,284	1,327	1,370	1,332	1,364	1,396	1,427	1,459
Distribution and retail services tariff (SLRs/kWh)	**2.10**	**2.08**	**2.07**	**2.06**	**2.03**	**2.88**	**2.86**	**2.85**	**2.83**	**2.82**

GWh = gigawatt-hour, SLRs = Sri Lanka rupees, SLRs/kWh = Sri Lanka rupee per kilowatt-hour.

Note: The above includes only the allowed costs for distribution and retail supply. Costs of energy purchased from transmission is passed through to customers.

Sources: Public Utilities Commission of Sri Lanka. Decision document on electricity tariff 2011, July 2011; Decision on Revenue Caps and Bulk Supply Tariffs 2016–2020, February 2016.

Ever since the electricity tariff methodology was streamlined in 2011, in real terms, LECO distribution and retail services cost per kWh has declined, though marginally, indicating the improving financial efficiency of the company.

After initial adjustments over 2011–2013, LECO profits from distribution have now stabilized.

The first few rounds of the new tariff methodology were based on a number of assumptions. As the methodology matured, the profits of LECO stabilized. From the business of electricity distribution and supply, LECO has maintained remarkably good profits since 2011.

LECO's operational asset base is SLRs10 billion and the return on assets is limited to 2%. Therefore, the allowed profit from electricity distribution business is estimated to be limited to about SLRs250 million per year, by 2018. The company worked within the allowed revenue and earned profits in excess of the 2% return on assets allowed, in 2017, 2018 and 2019. LECO's corporate profits are significantly more than the profits from distribution business. The remaining profits are dividends earned on investments and interest earned on deposits maintained at banks.

From a loss-making utility from its inception until 1990, the company has earned increasing profits. This gradual increase peaked in 2006 with a net profit before tax of SLRs965 million. Thereafter,

the profit before tax declined owing to unfavorable bulk supply tariffs. LECO reported losses in 2009 and 2010 (see figure 4).

LECO has not borrowed by way of long-term loans after the three loans secured from ADB in the initial years over 1980–1990 to rebuild the distribution network. These loans have been fully settled by the end of 2013. However, the total liabilities of LECO have continued to increase year on year from SLRs4.6 billion at the end of 2000 to SLRs11.5 billion by the end of 2017 with the increase of differed income, trade payables, and employee retirement benefits. However, equity in LECO has increased nine times from SLRs3.4 billion at the end of 2000 to SLRs29.1 billion by the end of 2017, strengthening its balance sheet.

LECO now has predictable profits from distribution and retail supply activities. However, as the network assets are aging, LECO has to invest heavily in network upgrades and distribution automation.

The company plans to invest $50 million over 2022–2025 to improve reliability of its network. This project would almost double the asset base of LECO. Much of the planned investments would neither increase sales nor reduce losses; investments are to replace aging assets and to improve reliability of electricity supply through distribution automation.

Figure 4: Allowed Profits and Actual Profits of LECO's Distribution and Supply Activities

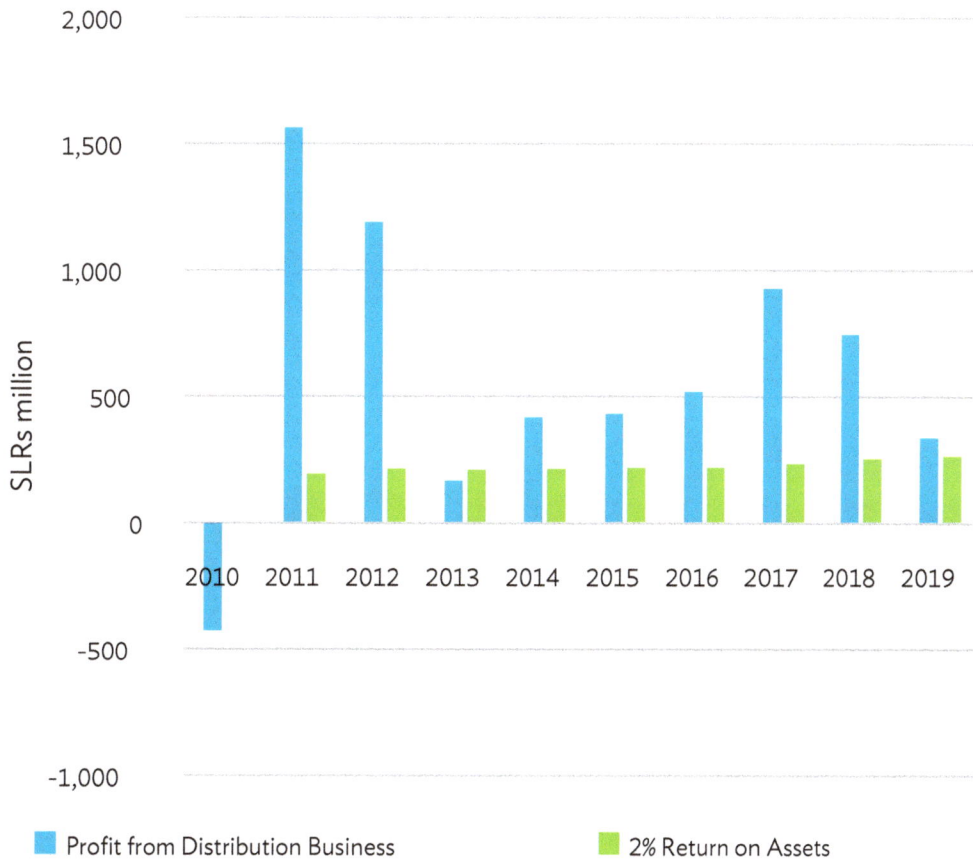

Legend:
- Profit from Distribution Business
- 2% Return on Assets

SLRs = Sri Lanka rupees.

Sources:
Upto 2017 – From Audited Financial Statements of LECO and Annual Reports of LECO,
2018 and 2019 – From Regulatory Accounts of LECO.

Upcoming investments have received regulatory approval through the biennial submission and review of the distribution plan. As such, LECO expects the upcoming investments to be compensated through the distribution allowed revenue.

Upcoming investments would double the asset base of LECO and raise the reliability of supply.

However, the distribution revenue currently stands at 3.20 SLRs/kWh and is estimated to increase below the inflation rate. LECO expects to implement efficient operation and maintenance practices to offset the relatively large debt repayments and interest expenses, once the upcoming loan for distribution upgrades has been disbursed.

LECO Operations 2020

7 branch offices

23 customer service centers

2,546 distribution substations

1,887 bulk supply substations

1,034 km of 11 kV distribution lines

3,806 km of 400/230 V distribution lines

1,580 GWh sales to customers

66 GWh purchased from rooftop solar PV

8,507 Customers with rooftop solar PV

600,729 customers

1,521 employees

LECO distribution network (LECO file photo).

A LECO customer with a net-metered rooftop solar unit. LECO offers net-metered distributed generation since 2008 (Photo by Harsha Wickramasinghe, LECO customer).

Gearing up for the Future

The Company Ventures into Other Businesses

LECO, fulfilling the true nature of a private sector entity, ventured into other businesses. LECO's articles of association allows the company to invest in other ventures and LECO has active investments in the following:

(i)   Ante LECO Metering Company (Pvt) Limited: 70% equity share

(ii) West Coast Power (Pvt) Limited: an equity partner with a 18% stake

(iii) LECO Projects (Pvt) Limited: a fully owned subsidiary

Ante LECO Metering Company is the only energy meter manufacturing facility in Sri Lanka, established by LECO in a joint venture with Ante Meter Company of the People's Republic of China. The company manufactures all energy

Electronic meters being tested at **ANTE LECO Metering Company (Pvt) Ltd.** All new customers in Sri Lanka receive a LECO–made meter (Photo by Bhasara Sirisinghe).

Sri Lanka imported electricity meters until 2018. Now LECO manufactures the full requirement its factory in Sri Lanka.

meters for household customers of both CEB and LECO. The company recently expanded the manufacturing facility to produce high-quality smart meters. Meters are tested and certified at KEMA laboratory, The Netherlands, with complete type testing according to standards International Electrotechnical Commission (IEC) 62052-11 and IEC 62053-21.

The company provides meter testing and calibration facilities to the highest accuracy and holds International Organization for Standardization (ISO) 9001:2008 and ISO IEC 17025:2005 certificates for its quality service.

West Coast Power is an independent power producer, with a capacity of 300 MW and providing electricity to the national grid. The company is owned by several shareholders in which LECO owns 18% stake.

LECO Projects provides a spectrum of services for LECO as well as for LECO clients ranging from electricity distribution projects to beyond-the-meter services.

Support to Accelerate Deployment of Rooftop Solar Power

LECO commenced promoting rooftop solar photovoltaic (PV) connections since 2008. By the end of 2020, LECO reached 8,507 customers with 50 MW of connected solar PV capacity compared with CEB's 20,341 customers with 275 MW of solar rooftop installations.

To promote rooftop solar PV among household customers, LECO offered a financing package, a loan of SLRs1.5 million, amounting to 75% of the solar PV system supplier's costs, repayable over 7 years at an interest rate of 8% per year, when commercial borrowing rates were 12%. LECO is also participating in the ADB-financed Rooftop Solar Power Generation Project since 2018.[8] The project aims to disburse $50 million through commercial banks to build about 60 MW of rooftop solar PV capacity at customer premises.

LECO improved the connection method of rooftop solar PV by installing a production meter, in addition to the two-way export-import meter. The two-way meter records only the net of imports and exports between the customer and the grid. It is not possible to record the actual amount of electricity produced and used by the customer because such usage is direct. The new smart meter comes with the capability to capture the full production, import and export information, thus assisting to gauge the energy performance of the systems installed. Additionally, it helps to keep the national energy accounts correct.

LECO was the first distribution utility to connect net-metered rooftop solar PV systems in Sri Lanka.

LECO is already experiencing power quality issues owing to very high daytime electricity injection to low-voltage distribution lines from solar PV. Until 2016, only net-metered solar PV systems were allowed. Two extended schemes are now offered, a "net-accounting" scheme where surplus energy

---

8 ADB 2017. Sri Lanka Solar Power Generation Project, ADB, Manila. https://www.adb.org/projects/documents/sri-50373-002-rrp

LECO has to either curtail hosting the rooftop solar PV program in specific areas, introduce a voltage drooping scheme, or instigate inverter dropout.

from solar PV is paid for, and the "net-plus" scheme where rooftop solar PV is treated as a stand-alone microgenerator.

Higher capacity additions and higher reverse power flows toward the grid from the customer end are now visible at midday. In some residential areas, reverse power flow during daytime is significantly more than the daytime demand. By 2019, LECO

had recorded 12 such specific cases of overvoltage in the low-voltage distribution network. The problem is expected to grow.

Solving Emerging Problems of Distributed Generation

LECO purchased 66 gigawatt-hours (GWh) from rooftop solar PV in 2020, amounting to 4% of total purchases. LECO wants to minimize curtailing rooftop solar PV owing to network constraints.

Since 2018, LECO has been assisting a joint research project at the University of Moratuwa, Sri Lanka and the University of Wollongong,

Machinery and skills. Having required machinery and skilled staff is the story behind LECO's success (Photo by M. A. Pushpa Kumara).

Australia. The study objective is to develop solutions to the problem of rising customer-end voltages and the broader issues of determining the "hosting capacity" (capacity to accommodate rooftop solar PV capacity).

The ongoing research has established rules for accommodating a rooftop solar PV system, on the basis of line lengths, customer mix, and potential random location of applicants to connect solar PV system.

LECO is also planning research on possible options to increase the hosting capacity of low-voltage networks, to facilitate more distributed generation.

Upgrades for Improved Reliability

Early assets of LECO in areas of councils taken over in stage 1 and rebuilt by 1987 have now come of age. Thirty-five years of service is a good time to review and replace assets, and to address problems caused by aging, the ever-changing built-environment, and customer expectations. Technology has advanced and distribution automation is the next logical improvement to serve customers better.

LECO planned to progressively improve the reliability of supply, but the progress is not satisfactory. Key challenges are much of the distribution assets are over 30 years old, customer density has increased beyond the forecast levels, and space available to expand the distribution network is limited.

In 2012, unplanned outage duration of the LECO network alone was 9.1 hours per customer, which had risen to 21.7 hours per customer in 2019. The increase had been persistent over 2016–2019. Outage duration owing to network upstream of LECO and planned outages have both slightly

LECO uses 11 kV aerial bundled conductors to get through congested urban spaces, but their capacity will soon be exhausted. An 11 kV transition point from bare wires to insulated wires (Photo by Thisara Karunathilake).

declined over the same period. Therefore, LECO has the challenge to progressively reduce the total outage duration to less than 10 hours per year per customer, as expected of a modern utility.

Most unplanned outages are caused by human activity, animals, and weather-related events. The duration of such events has increased since 2016 due to the policy of not switching on 11 kV lines after a tripping event until the line is physically inspected. Even if this issue is solved, the limitation of lines and vulnerability to external events will remain as long as the LECO medium voltage network remains uninsulated and overhead.

ADB is assisting LECO to implement the best strategy to upgrade the medium voltage network. An upcoming proposed project by ADB would provide a new grid substation at Kelaniya, a town to the east of the capital Colombo, and lay new 33 kV underground distribution lines to improve the upstream reliability of the LECO network (see Figure 5). Therefore, the network topology is proposed to be changed. Currently, it is an entirely overhead, uninsulated network of upstream 33 kV lines of CEB and LECO's own 11 kV lines. The new topology will use the advantages of direct access to a grid substation, and build new upstream underground 33 kV lines and some new 11 kV underground lines.

LECO does not see the merit in changing the entire distribution network from overhead to underground.

In the upcoming upgrades, LECO plans to install "packaged substations." These ensure the required insulation against impacts of weather and human activity while being cost-effective. Therefore, the LECO system will have new assets, of specifications not

The existing and new switchgear would all be fully automated, participating in distribution automation, to reduce outages and their duration.

previously used, such as (i) outdoor 33 kV for low-voltage substations to serve bulk customers or retail customers, (ii) 33 kV underground lines, and (iii) 11 kV for low-voltage packaged substations.

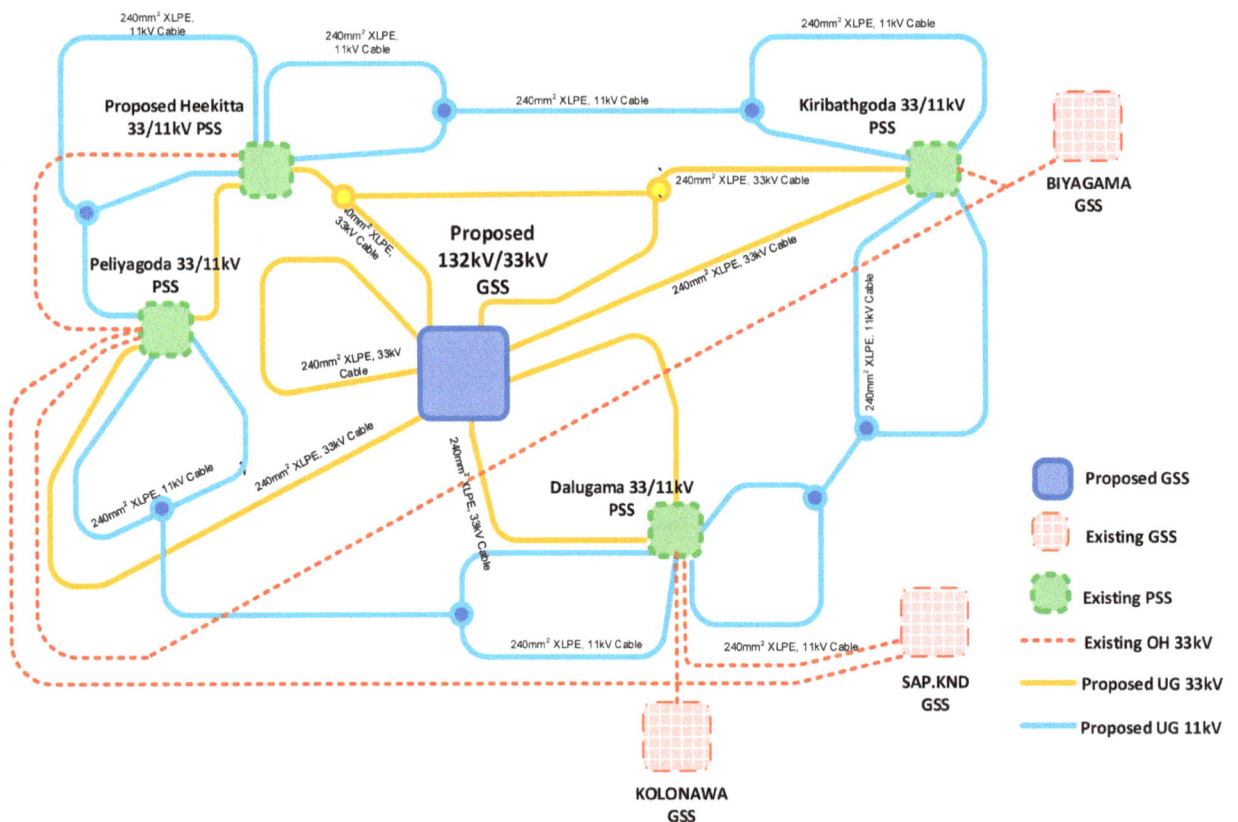

Figure 5: Planned Upgrades to the LECO Network in Kelaniya

GSS = grid substation, kV = kilovolt, mm = millimeter, PSS = primary substation, UG = underground, XLPE = cross linked polyethylene cable.
Source: LECO Engineering Division.

Similar upgrades to underground 11 kV and 33 kV lines are planned for several other areas of LECO operations and included in the upcoming project being considered by ADB for financing. LECO is presenting distribution automation activities too, with an automated assessment of reliability indexes by linking the distribution control center with the reliability database.

The Next Move: Distributed Generation in Smart Grids

LECO is gearing to transform itself from a traditional distribution utility serving customers to a modern utility that provides a pathway for energy exchange between customers and the associated ancillary services.

Power flows on distribution lines will be increasingly bidirectional, with storage available with the utility service providers as well as with customers. Electricity tariffs would have a new meaning, where the utility service price and the optional sources available to the customer (battery storage, renewable-based generation, vehicle-to-grid supply, or energy credits held by the customer) would compete or complement each other. Technical and commercial solutions would be many and diverse.

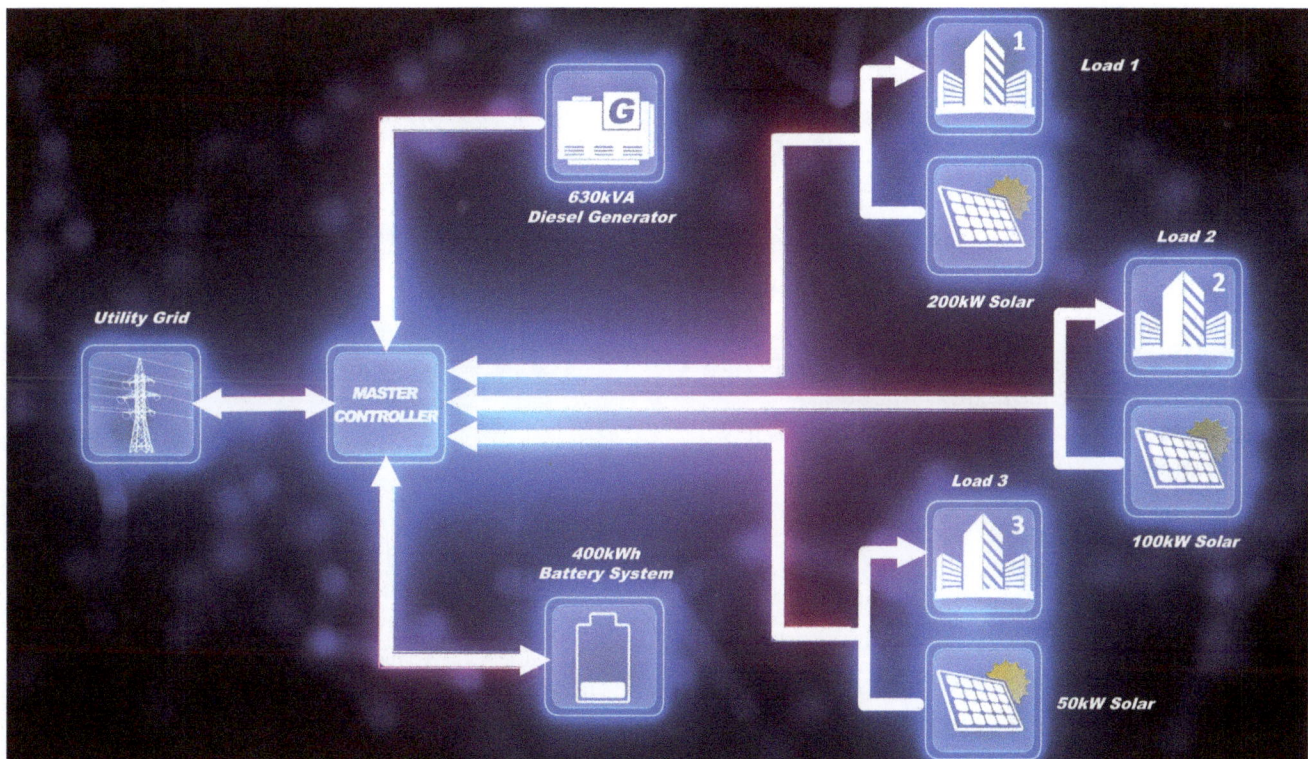

Schematic diagram of the ADB-assisted microgrid at University of Moratuwa - The utility power, solar PV, batteries and a diesel generator are integrated through an intelligent controller, to deliver technically and commercially efficient power to a section of the University. The microgrid, to be fully operational from 2022, will provide exposure to students to state-of-the-art technology, and facilitate graduate studies and research.

To demonstrate such technological and commercial issues and solutions, LECO with ADB pioneered a proposal to establish a live microgrid.[9]

The microgrid project using a grant of $1.8 million from the ADB-administrated Clean Energy Financing Partnership Facility at the University of Moratuwa has been implemented. The microgrid uses three rooftop solar PV systems, a central battery storage, and the LECO grid, all working in harmony to provide a technically and commercially optimal electricity supply to a section of the university, serving a peak demand of about 300 kW.

> In 2018, LECO with ADB launched research into microgrids and Sri Lanka's first pilot microgrid has been established at the University of Moratuwa.
>
> The system is to be used as a live demonstration of its technical and commercial capabilities, as well as the limitations of renewable energy microgrids.

It will enable staff and student research into further optimizing the concept, looking forward to Sri Lanka's future grid where distributed generation is likely to be a significant contributor for generation.

The project includes a direct current service network as well. This will avoid the use of inverters and rectifiers to serve computer equipment in the microgrid service area, which will be operated on direct current straight from the solar PV system or the battery storage.

Financial Independence

LECO actively participated in the ADB-funded exercise over 2010–2011 to establish regulatory instruments for a distribution entity to perform under the law. These instruments included tariff methodology, allowed charges, distribution code, supply services code, and performance standards. Subsequent regulations issued on demand-side management and performance standards have also been complied with.

LECO has now established the procedure to measure and report power quality and reliability. LECO provides all information to the regulator, for ex-ante tariff calculations and ex-post adjustments. Major elements of reforms have yet to be implemented in the upstream distribution and transmission network managed by CEB.

> The Electricity Act of 2009 and its implementation over the past 10 years have improved transparency and LECO's technical and financial performance.
>
> Full implementation of unfinished reforms and regulatory improvements would strengthen Sri Lanka's electricity industry to reach its goals of efficiency, reliability, and sustainability.

While LECO's financial operations and regulatory accounts are clearer, the upstream bulk–supply entity–CEB transmission has been operating at a loss ever since the new tariff methodology was established.

---

[9]   ADB. 2016. Supporting Electricity Supply Reliability Improvement Project- Grant 0486-SRI: Renewable Energy Micro-grid, ADB, Manila. https://www.adb.org/projects/documents/grant-0486-sri-sesrip-grj

The Public Utilities Commission of Sri Lanka, the regulator, has not been able to compensate CEB for the subsidies awarded to customers, some of which have flowed through LECO to customers. However, LECO has been awarded its full allowed revenue, resulting in financial independence and consistent profits since the introduction of the tariff methodology.

Past and Emerging Challenges

LECO was established as an independent entity, only answerable to its shareholders. That LECO's shareholders are themselves government entities did not cause LECO to be different to any other company operating under the Companies Act. The sector experience, engineering knowledge, leadership, and commitment of the chairman/men, supported by the equally committed general manager and staff in the formative years, significantly contributed to LECO's success with internal reforms as well as its being an example to the regional utility industry.

Senior managers developed planning, construction, operation, and customer service procedures; and experienced results and were allowed to implement corrective actions.

> Senior management staff were allowed to remain in key positions for a sufficiently long duration.

Erosion of administrative independence commenced with the appointment of a chairman/men in 2001 who did not have the necessary expertise to chair the premier institution. At the time, LECO was fast becoming an example to distribution utility industry in South Asia.

External interference caused a mass exodus of senior managers leaving the company, causing a vacuum in middle and senior management. Nevertheless, middle management has been able to hold the company together, weathering many external shocks, while continuing the journey toward excellence in distribution engineering and customer service.

Under the tariff methodology established by the Electricity Act of 2009, LECO will continue to report healthy distribution revenue and profits, which is considered a very positive outcome of the regulatory environment that has been created in Sri Lanka.

Three-phase meter manufactured at ANTE LECO Metering Company (Pvt) Ltd, ready to be deployed after testing (Photo by Bhasara Sirisinghe).

LECO's ability to respond to emerging challenges and withstand these external pressures, in turn, depends on the quality, strength, commitment, qualifications, and experience of the company's board of directors and its workforce.

The greatest value LECO added to the utility industry, as a state-owned entity run with the efficiency of most private entities, could not be replicated elsewhere in the electric utility industry.

The "LECO model" still remains alive, active, and relevant—ready to be replicated in other utilities in Sri Lanka and elsewhere, and ready to prosper in the future, maneuvering among unforeseen challenges.

It should be remembered that many of LECO's technical innovations, which were considered by CEB to be unnecessary or too expensive in the 1980s and 1990s, have since been adopted by CEB.

LECO microgrid project at University of Moratuwa, Sri Lanka, uses 350 kW of solar PV placed on multiple university buildings, as the main source of energy (Photo by Tharindu de Silva).

Vision and Strategy into the Future

Narendra de Silva, General Manager, LECO

Appointed general manager and chief executive officer of LECO in 2020, Narendra de Silva had worked in LECO since 2007. He was the Engineering, Research and Development manager from 2007 to 2012 and head of Engineering from 2012, until he was appointed as general manager. He holds a 1st Class honors degree in electrical engineering from University of Moratuwa, Sri Lanka. He earned his PhD from Heriot-Watt University, United Kingdom.

Recounting his 14 years of work in LECO, Narendra said that LECO continued with the company's strategy of providing a higher-quality service to customers. "LECO pioneered the implementation of the net metering scheme for rooftop solar PV in 2008. In 2020, we purchased 66 GWh of energy from customers with rooftop solar PV. We, as an urban distribution utility serving about 155 customers per km of low-voltage lines, are already facing daytime congestion in low-voltage distribution networks, caused by rooftop solar PV. We are actively working on solutions to the problem," he said.

When asked what the greatest challenges to LECO were at this time when the company is approaching the end of the fourth decade of its existence, Narendra stated, "Our network requires investments to overcome the effects of aging and obsolescence. Customers expect an automated distribution control system, whereas we still require operator intervention when network segments trip out. Our medium-voltage network is largely uninsulated, requiring care to keep the vegetation under control. Therefore, improving the reliability of electricity supply to match the expectations of the modern society is our first priority. Accordingly, we are planning a major investment program for implementation over 2022–2025."

He emphasized the requirement for higher observability of the distribution network and the utility functions in real-time for the closer control. "This is a must for a modern distribution utility, constrained by the ever-increasing challenges such as renewable admission, and striving to satisfy the customer expectations of the modern society. Our Smart Grid equipped with GIS-enabled Internet of Things devices, connected switches, smart meters, and network monitoring devices are expected to deliver us a more transparent and observable system. Through this, we expect to enhance our services while reducing costs and environmental impacts," he said.

Narendra continued to describe the second key challenge, which is to accommodate distributed generation, technically and commercially affecting LECO's operations. He explained the gaps in regulatory practice, denying LECO the incremental costs of purchasing energy from distributed generation at prices above the bulk supply tariff, while being required to invest in online upgrades to accommodate such generation. "LECO is concerned that economic regulation is not functioning as required by the law, although LECO is pleased that certain regulatory functions are active. LECO's expenses on bulk purchases from the national grid are now largely predictable, which is useful, especially when our selling prices continue to be predefined."

"Thanks to the regulatory reforms over 2010–2011 supported by ADB's technical assistance, LECO's profits are now predictable, which encourages the management and staff to cautiously adhere to approved costs and watch the network losses. Our allowed network loss was 4.23% for year 2019, and we achieved 3.98%. If the losses increase to 5% against the regulatory limit of 4.23%, that will wipe out almost the entire annual return on investment allowed in the tariff," he said.

"Gearing LECO to meet the ever-changing role of a distribution utility is our third challenge," he said. He explained that distribution utilities are increasingly becoming service providers to facilitate energy exchange between suppliers and users. In Sri Lanka, LECO is both the distributor and the supplier in the service area. The customer has no choice on who his supplier would be. Electricity wheeling is still not available. He explained that major improvements are required to regulatory practices and how subsidies are managed before Sri Lanka launches a truly liberal electricity industry. "LECO will be happy to lead Sri Lanka's distribution industry toward a liberalized electricity market once the correct regulations and their implementation on the ground are demonstrated by the government and the Public Utilities Commission," he concluded.

References

ADB. 1984. *Report and Recommendation of the President to the Board of Directors: Proposed Loan to the Government of Sri Lanka for the Secondary Towns Power Distribution Project.* Manila.

ADB. 1984. *TA 578-SRI, Final Report Prepared by Beca Worley International New Zealand in Association with Personnel Administration of New Zealand.* Manila.

ADB. 1990. *Completion Report: Secondary Towns Power Distribution Project in Sri Lanka.* Manila.

ADB. 1994. *Completion Report: Secondary Towns Power Distribution Project II in Sri Lanka.* Manila.

ADB. 2012. *Technical Assistance to Sri Lanka for Capacity Building for Power Sector Regulation.* Manila.

ADB. 2016. *Report and Recommendation of the President to the Board of Directors: Proposed Loan and Administration of Grant to the Democratic Socialist Republic of Sri Lanka for Supporting Electricity Supply Reliability Improvement Project.* Manila.

ADB. 2017. *Technical Assistance to Sri Lanka for Power System Reliability Strengthening Project.* Colombo.

Government of Sri Lanka, Ministry of Power and Energy. 1982. *Report of the Committee on Improvement of Electricity Distribution in Local Authority Areas.* Colombo.

Government of Sri Lanka, Ministry of Power and Renewable Energy. 2016. *Electricity (Safety, Quality and Continuity) Regulations, Gazette No. 1975/44.* Colombo.

Lanka Electricity Company (Pvt) Ltd. 1989. *Institutional Strategy Study for Lanka Electricity Company Limited, UCI international.* Colombo.

Lanka Electricity Company (Pvt) Ltd. 1990. *Annual Reports 1990 to 2018.* Colombo.

Lanka Electricity Company (Pvt) Ltd. 2019. *Medium Voltage Development Plan 2019–2028.* Colombo.

Lanka Electricity Company (Pvt) Ltd. 2019. *Proposal for LECO Distribution System Reliability Strengthening Project (DSRSP), System Development Department.* Colombo.

Lanka Electricity Company (Pvt) Ltd. 2021. *Statistical Digest 2020.* Colombo.

Lanka Electricity Company (Pvt) Ltd. *Corporate Plan 2018–2022.* Colombo.

66

Public Utilities Commission of Sri Lanka. 2010. *Decision on Electricity Tariffs (Effective from 1 January 2011)*. Colombo.

Public Utilities Commission of Sri Lanka. 2016. *Decision on Revenue Caps and Bulk Supply Tariffs 2016–2020*. Colombo.

Public Utilities Commission of Sri Lanka, Sri Lanka Energy Managers Association. 2016. *Revision of Network Loss Targets*. Colombo.

Sri Lanka Sustainable Energy Authority. 2019. *Sri Lanka Energy Balance 2018*. Colombo.

World Bank. 1988. *Integrated National Energy Planning in Sri Lanka, Mohan Munasinghe and Peter Meier*, section 3.2, pp. 54–60. Washington, DC.

www.ingramcontent.com/pod-product-compliance
Lightning Source LLC
Chambersburg PA
CBHW050051220326
41599CB00045B/7364